BASIC MATHEMATICAL ANALYSIS: THE FACTS

IAN S. MURPHY

Department of Mathematics
University of Glasgow

ARKLAY PUBLISHERS

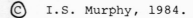
ISBN   0 9507126 2 0

First Edition 1980.
Second Edition (with corrections) 1984.

Published by Arklay Publishers,
64 Murray Place, Stirling, Scotland.   FK8 2BX.

Printed and Bound in Great Britain by
Bell and Bain Ltd., Glasgow

PREFACE

This book is intended as an aid for students attending a basic course on Mathematical Analysis. Most students embarking on such a course already have some working knowledge of Calculus and therefore this book makes little attempt to develop further technical skill in differentiation and integration. Instead its main aim is to provide a clear overall picture of the development and properties of the exponential, trigonometric and hyperbolic functions: often this area of work is not fully appreciated in a first Calculus course. Concepts like continuity, differentiability, compactness, infinite series and integration arise naturally in the achievement of this aim. The definition of the exponential is the key to the whole matter and there are various possible approaches as mentioned in §208. In this book the decision has been to approach via power series and this has dictated the overall plan. Much of the text is however independent of the particular approach being followed. There are a few places where, in an attempt to widen the reader's experience or for reasons of convenience, I have anticipated ideas not defined till later; the underlying structure is however unaffected.

Many students have made comments to me about the style and layout of mathematical books in general. I have written this book with these comments in mind.

The book is based on lectures given over the past few years to students setting out on an Honours course in Mathematics at the University of Glasgow. The treatment has been influenced by the book of Professor R.A.Rankin mentioned on page 221. I am also indebted to Professor Rankin for several helpful comments on the completed text of this book and to Dr. Ian Anderson, who very kindly read the completed text in detail and made many suggestions for improvement. My thanks also go to Mrs. Maureen McGrady for her excellent typing of a substantial part of the book.

Ian S.Murphy

University of Glasgow.

July, 1980

CONTENTS

I. BASIC IDEAS

1. In this chapter we shall work in R, the set of real numbers. Bounds for subsets of R, inequalities and estimation are our main concern but we shall also look at some points of technique in connection with functions, counterexamples and quantifiers.

2. THE SET OF REAL NUMBERS AS AN ORDERED FIELD

We take the set R of real numbers as known and we assume that R is a field with the usual operations of addition and multiplication. This means in particular that we can assume, for non-zero $a \in R$ and positive integers m, n, that

$$a^m a^n = a^{m+n}, \qquad (a^m)^n = a^{mn}, \qquad a^0 = 1.$$

We define $0^0 = 1$.

We also take the natural ordering $>$ on R as known and we assume that $>$ satisfies the following conditions.

1. For every pair $x, y \in R$, exactly one of the statements $x > y$, $x = y$, $y > x$ is true.

2. The statements $x > y$ and $x - y > 0$ are equivalent.

3. If $x > 0$ and $y > 0$, then $x + y > 0$ and $xy > 0$.

The statement $x > y$ is read as "x is greater than y" and a number x is <u>positive</u> if $x > 0$.

It is worth noticing in the shadow of the above assumptions that in a general field there may be no such ordering satisfying the conditions 1,2,3. A case in point is the field Z_5 consisting of the set $\{0,1,2,3,4\}$ with addition and multiplication modulo 5; this is the subject of Ex.1.33.

3. N, Z and Q are the sets of positive integers, all integers and rational numbers. Rational numbers are those real numbers expressible as quotients of two integers, e.g. $-\frac{1}{2}$. This representation as p/q is unique if we demand that $q > 0$ and that p and q have no factor in common apart from 1.

The members of $R - Q$ are the irrational numbers. Notice that $\sqrt{2}$ is irrational as Ex.1.32 shows. Notice also that no matter how close together two real numbers x, y may be, we can still find both rational and irrational numbers between them. To see this suppose without loss of generality that x,y are both positive. Then mark off successive rational points $1/n$, $2/n$, $3/n$, ... or successive irrational points $\sqrt{2}/n$, $2\sqrt{2}/n$, $3\sqrt{2}/n$,

... (n$\in$N) on the real line. By taking n sufficiently large we can reduce the distance between successive points in each sequence sufficiently to make at least one point in each sequence lie between x and y.

We intend to assume the existence of rational powers of positive real numbers, e.g. $\sqrt{2}$, right from the start of this book. Their existence is actually proved in §90 but it is convenient to have them available from the outset. Later on in §203 and §210 we actually go further and assign a meaning to irrational powers.

4. For real numbers x and y, x $\geq$ y means that x is greater than or equal to y. So the statements 7 $\geq$ 5 and 7 $\geq$ 7 are both true. Moreover, if A = {1, 5, 6, 8}, then the statement that a $\geq$ 0 for all a $\in$ A is also true. The statement x < y is equivalent to y > x and is read as x is less than y. Similarly x $\leq$ y is equivalent to y $\geq$ x.

The statements that x is positive, negative and non-negative mean respectively that x > 0, x < 0 and x $\geq$ 0. Also, for a, b, c $\in$ R, notice that b lies between a and c means that a $\leq$ b $\leq$ c. Notice also the following notation for intervals in R:

[a,b] = {x: a $\leq$ x $\leq$ b},   ]a,b[= {x: a < x < b},

]a,b] = {x: a < x $\leq$ b},   [a,∞[= {x: x $\geq$ a}.

5. The modulus of x$\in$R (denoted $|x|$) is x if x $\geq$ 0, and is -x otherwise. As such the modulus is non-negative, and $|ab| = |a||b|$. It is often helpful to think of $|x-y|$ as the distance between the points x and y on the real line. Notice also the occasionally useful expressions for the maximum and minimum of two real numbers x, y $\in$ R:

$$\max(x,y) = \tfrac{1}{2}(x+y+|x-y|), \quad \min(x,y) = \tfrac{1}{2}(x+y-|x-y|).$$

Notice also that, for all x, y $\in$ R and for each n$\in$N,

$$x^n - y^n = (x - y)(x^{n-1} + x^{n-2}y + \ldots + y^{n-1}).$$

Similar factors for $x^n + y^n$ can only be found in cases where n is odd. In particular,

$$x^3 + y^3 = (x + y)(x^2 - xy + y^2).$$

INEQUALITIES

6. As a consequence of the properties 1,2,3 of the ordering > detailed in §2, we can transpose terms in an inequality just as in an equation. For example, x > y + z if and only if x - y > z.

A further consequence of these properties is the following result about the multiplication of an inequality by a real number.

RESULT. It is permissible to multiply both sides of an inequality by a positive number. If both sides are multiplied by a negative number, then $>$, $\geq$, $<$, $\leq$ must be changed to $<$, $\leq$, $>$, $\geq$ respectively.

Proof. We prove two cases to illustrate.

Firstly, suppose $a > b$ and $c > 0$. Then, from properties 2 & 3 of §2, it follows in turn that $a - b > 0$ and $c(a - b) > 0$. So $ca - cb > 0$, i.e. $ca > cb$, by property 2.

Secondly, suppose $a < b$ and $c < 0$. Then, $b > a$ and $0 > c$. So $b - a > 0$ and $-c > 0$. So, by the first part, $-c(b - a) > 0$, i.e. $-cb + ca > 0$, i.e. $ca > cb$.

7. Notice that $|a| \leq 6$ is equivalent to $-6 \leq a \leq 6$. The following example uses this idea.

EXAMPLE. Suppose that $|x - a| < 1$. Prove that $a - 1 < x < a + 1$.

Solution. Since $|x - a| < 1$, $-1 < x - a < 1$. Transpose on each side to get $a - 1 < x < a + 1$.

EXAMPLE. Suppose that $|x - a| < \frac{1}{2}a$. Prove that $a/2 < x < 3a/2$.

8. THE TRIANGLE INEQUALITY

RESULT. For all $a, b \varepsilon R$, $|a+b| \leq |a| + |b|$.

Proof. Notice that $-|a| \leq a \leq |a|$
and $-|b| \leq b \leq |b|$.
On addition get $-(|a| + |b|) \leq a+b \leq (|a| + |b|)$.
So from the idea of §7 conclude that $|a+b| \leq |a| + |b|$.

EXAMPLE. Show that if $|x| \leq 1$, then $|x^5 - 12x + 85| \geq 72$.

Solution. $85 = |x^5 - 12x + 85 + 12x - x^5|$
$\leq |x^5 - 12x + 85| + |12x - x^5|$
$\leq |x^5 - 12x + 85| + |12x| + |x^5|$
$\leq |x^5 - 12x + 85| + 12 + 1$ if $|x| \leq 1$.

On transposing find that
$72 \leq |x^5 - 12x + 85|$ if $|x| \leq 1$ as required.

(Compare §11(i).)

9. The following secondary form of the triangle inequality also merits some attention.

RESULT. For all $x, y \in R$, $||x| - |y|| \le |x - y|$.

Proof. Put $a = x - y$ and $b = y$ in the triangle inequality. So $|x - y + y| \le |x - y| + |y|$,

i.e. $|x| - |y| \le |x - y|$. ...(1)

Then put $a = y - x$ and $b = x$ in the triangle inequality to get $|y - x + x| \le |y - x| + |x|$,

i.e. $-|x - y| \le |x| - |y|$. ...(2)

Then, from (1) and (2),

$$-|x - y| \le |x| - |y| \le |x - y|.$$

So, using the idea of §7 we deduce the required result.

EXAMPLE. It is given that $|x - a| < \frac{1}{2}|a|$. Prove that $|a|/2 < |x| < 3|a|/2$.

Solution. Use the above result to see that

$$||x| - |a|| \le |x - a| < \frac{1}{2}|a|.$$

Then use §7 to conclude that $-\frac{1}{2}|a| < |x| - |a| < \frac{1}{2}|a|$.

Transposition of terms gives $\frac{1}{2}|a| < |x| < \frac{3}{2}|a|$, as asked.

ESTIMATION

10. It is often necessary to estimate an expression, i.e. to find an inequality relating the given expression to something simpler. The triangle inequality and the techniques of §11 and §12 are very useful in this connection.

11. (i) Notice first that $y \ge -|y|$ for all $y \in R$. So, for example,

$$85 + 12x - x^5 \ge 85 - 12|x| - |x|^5$$
$$\ge 85 - 24 - 32 \quad \text{for} \quad |x| \le 2,$$
$$\ge 29 \quad \text{for} \quad |x| \le 2.$$

(ii) Notice also that a fraction (with numerator and denominator positive) can be increased by increasing the numerator and/or decreasing the denominator and can be decreased by decreasing the numerator and/or increasing the denominator. For example,

$$\frac{(x-1)}{(x-4)(x+9)} \ge \frac{3}{14(x-4)} \quad \text{for all} \quad x \in]4,5[,$$

because $(x-1)$ exceeds 3 and $(x+9)$ is less than 14 for all points $x \in]4,5[$.

12. THE DOMINANCE OF THE TERM OF HIGHEST DEGREE IN A POLYNOMIAL FUNCTION

For large values of x, the behaviour of a polynomial like x^2+9x+5 is decided by the behaviour of its dominant term. The following result (for a polynomial of degree 2) and its corollary (for a general polynomial of degree n) make this idea precise.

RESULT. Let a, b, $c \in R$ with $a > 0$. Then there exists a non-negative real number N such that

$$\tfrac{1}{2}ax^2 \le ax^2+bx+c \le \tfrac{3}{2}ax^2 \quad \text{for all} \quad x \ge N.$$

Proof. Notice that $\quad |bx| \le \tfrac{1}{4}ax^2 \quad \forall x \ge 4|b|/a \quad (=L)$

and $\quad |c| \le \tfrac{1}{4}ax^2 \quad \forall x \ge \sqrt{(4|c|/a)} \quad (=M).$

It follows that $\quad |bx|+|c| \le \tfrac{1}{2}ax^2$ and $-\tfrac{1}{2}ax^2 \le -|bx|-|c|$

$$\forall x \ge N = \max(L,M).$$

Then since

$$ax^2-|bx|-|c| \le ax^2+bx+c \le ax^2+|bx|+|c| \quad \text{for all} \quad x \in R,$$

it follows that

$$ax^2 - \tfrac{1}{2}ax^2 \le ax^2+bx+c \le ax^2 + \tfrac{1}{2}ax^2 \quad \forall x \ge N,$$

i.e. $\qquad \tfrac{1}{2}ax^2 \le ax^2+bx+c \le \tfrac{3}{2}ax^2 \qquad \forall x \ge N.$

COROLLARY. Let a_0, a_1, $\ldots$, $a_n \in R$ with $a_n > 0$. Then there exists a non-negative real number N such that

$$\tfrac{1}{2}a_n x^n \le a_n x^n + a_{n-1}x^{n-1}+\ldots+a_1 x+a_0 \le \tfrac{3}{2}a_n x^n \quad \forall x \ge N.$$

Proof. This is analogous to the proof of the above result: aim to make each of $|a_{n-1}x^{n-1}|$, $|a_{n-2}x^{n-2}|$, $\ldots$, $|a_0|$ less than $\tfrac{1}{2n}a_n x^n$ for sufficiently large x.

13. EXAMPLE.
In each of the following cases find a constant K such that the inequality holds for all positive integers n greater than or equal to some threshold value X:

(i) $\quad \left| \dfrac{n+6}{n^2+2} + \dfrac{(-1)^n}{n+5} \right| \le \dfrac{K}{n}$,
(ii) $\quad \dfrac{3n+7}{n^2+5n+9} \le \dfrac{K}{n}$,

(iii) $\quad \dfrac{3n+7}{n^2-5n+9} \le \dfrac{K}{n}$,
(iv) $\quad \dfrac{5n^3-5n+1}{n^2-2n+9} \ge Kn.$

Solution.

(i)
$$\left| \frac{n+6}{n^2+2} + \frac{(-1)^n}{n+5} \right| \leq \frac{n+6}{n^2+2} + \frac{1}{n+5} \qquad \forall n \geq 1$$

$$\leq \frac{n+6n}{n^2} + \frac{1}{n} \qquad \forall n \geq 1$$

$$\leq \frac{8}{n} \qquad \forall n \geq 1.$$

(ii)
$$\frac{3n+7}{n^2+5n+9} \leq \frac{3n+7n}{n^2+5n+9} \qquad \forall n \geq 1$$

$$\leq \frac{10n}{n^2} \qquad \forall n \geq 1$$

$$\leq \frac{10}{n} \qquad \forall n \geq 1.$$

(iii)
$$\frac{3n+7}{n^2-5n+9} \leq \frac{3n+7n}{n^2-5n+9} \qquad \forall n \geq 1$$

$$\leq \frac{10n}{\frac{1}{2}n^2} \qquad \forall n \geq X \quad \text{(by §12)}$$

$$\leq \frac{20}{n} \qquad \forall n \geq X.$$

(iv)
$$\frac{5n^3-5n+1}{n^2-2n+9} \geq \frac{\frac{5}{2}n^3}{\frac{3}{2}n^2} \qquad \forall n \geq X \quad \text{(by §12)}$$

$$\geq \frac{5n}{3} \qquad \forall n \geq X.$$

Compare the treatment of the denominators in parts (ii) and (iii). Notice that in part (ii) $n^2+5n+9 > n^2$ for all $n \in N$ and so replacing n^2+5n+9 by n^2 does decrease the denominator. On the other hand in part (iii) notice that n^2-5n+9 is not greater than n^2 for any value of n greater than or equal to 2 and in attempting to decrease the denominator we use the result (from §12) that $n^2-5n+9 \geq \frac{1}{2}n^2$ for sufficiently large values of n.

In doing estimation be careful to preserve the character of the expression. Notice in the above example how in both numerator and denominator the highest power of n is the only one eventually retained in the estimate. In general this is a good rule to follow where large values of n are involved.

BOUNDED SETS OF REAL NUMBERS

14. SETS WITH A MAXIMUM ELEMENT

Every finite subset of R has a maximum element and a minimum element. For example, for $B = \{1,4,8,12\}$, max B = 12 and min B = 1. An infinite subset of R however need have no maximum and/or no minimum. For example the interval $]-\infty,3[$ has neither a maximum nor a minimum element. The number 3 does not qualify as its maximum element as it is not in A.

15. SETS THAT ARE BOUNDED ABOVE AND/OR BELOW

A subset A of R is <u>bounded above</u> if there is a real number M such that for all $a \epsilon A$, $a \leq M$. In quantifier notation A is <u>bounded above</u> if and only if

$$\exists M \text{ such that } \forall a \epsilon A, \quad a \leq M.$$

The number M is called an <u>upper bound for A.</u>
Similarly A is <u>bounded below</u> if and only if

$$\exists m \text{ such that } \forall a \epsilon A, \quad a \geq m.$$

The number m is called a <u>lower bound for A.</u>
For example the interval $]-\infty,3[$ is bounded above and 3, $3\frac{1}{2}$ and 8 are all upper bounds. On the other hand $]-\infty,3[$ is <u>not</u> bounded below and there is no lower bound.
A subset A of R is <u>bounded</u> if it is bounded both above and below. For example $]-\infty,3[$ is <u>not</u> bounded but $[1,6[$ is bounded.

16. THE SUPREMUM OF A SUBSET OF R

This idea generalises the idea of a maximum element. The supremum of the set $]-\infty,3[$ is 3.

<u>Definition</u>. A real number σ is called the <u>least upper bound</u> or <u>supremum</u> of the subset A of R <u>if and only if</u>

(1) $a \leq \sigma \quad \forall a \epsilon A,$

and (2) $\forall \sigma'$ less than σ, $\exists a \epsilon A$ such that $a > \sigma'$.

(<u>Notation</u>: $\sigma = \sup A$.)

Look at the case where $A =]-\infty,3[$. Here numbers less than 3, like 2.8 are disqualified from being σ by condition (1). Also numbers greater than 3, like $3\frac{1}{2}$, are disqualified from being σ by condition (2). For example suppose $3\frac{1}{2}$ were σ. Then take $\sigma' = 3.2$ and try to satisfy condition (2). There is no $a \epsilon A$ with $a > 3.2$.

The number 3 is σ.   For, firstly a ≤ 3 for all
aεA, and secondly whatever   σ'   less than 3 you care to
choose (e.g.   σ' = 2.9), there exists aεA   such that
a > σ' (e.g.   a = 2.92).

Notice the force of the quantifier ∀ in
condition (2).   Notice also, as mentioned in §33
that the order of the quantifiers ∀ and ∃ in
condition (2) cannot be altered.

17.   EXAMPLE.   Let A = {1,4,8,12}.   Verify that
sup A = 12.

Solution.   First notice that   a ≤ 12   ∀aεA.   Then
take   σ' < 12   and notice that we can find the element 12
in   A   such that   12 > σ'.   (In fact in this case this
same element 12 can play a, where   a > σ', no matter
what value less than 12 is playing σ'.)   So   σ = 12.

This example is evidence of the fact that for a
set   A   which has a maximum element, sup A = max A.

18.   INFIMUM

Definition.   The real number   τ   is the greatest lower
bound or infimum of a subset   A   of   R   if and only if
   (1)   τ ≤ a   ∀ aεA,
and (2)   ∀ τ'   greater than   τ,   ∃aεA   such that a < τ'.
(Notation: τ = inf A.)

For example, inf N = 1   and   inf]3,∞[= 3.

19.   Notice that sets like   [3,∞[and   N   have no
supremum.   The following assumption (on which this
book is based) gives a guarantee about the existence of
suprema and infima:

ASSUMPTION.   Every non-empty subset of   R   that is
bounded above has a supremum, and every non-empty subset
of   R   that is bounded below has an infimum.   (These
suprema and infima are real numbers.)

20.   Notice that if A ⊆ R and K is an upper bound
for A, then automatically sup A ≤ K, because
sup A is the least among all upper bounds for A.
Similarly, if L is a lower bound for A, then automatically inf A ≥ L, because inf A is the greatest
lower bound for A.

Notice how these ideas are used in the following
example.

EXAMPLE. Let A be a bounded subset of R and let B = {b: b = 3-a, aεA}. Prove that inf B = 3-sup A.

Solution. (Notice that the definition of B means that B consists of all possible numbers 3-a, where aεA.)

Denoting sup A by σ, notice that

$$a \leq \sigma \quad \text{for all} \quad a\varepsilon A.$$
$$\text{So} \quad -a \geq -\sigma \quad \text{for all} \quad a\varepsilon A.$$
$$\text{So} \quad 3-a \geq 3-\sigma \quad \text{for all} \quad a\varepsilon A.$$
$$\text{So} \quad b \geq 3-\sigma \quad \text{for all} \quad b\varepsilon B.$$

So 3-σ is a lower bound for B and we conclude (by the above remark) that inf B $\geq$ 3-σ. So either inf B = 3-σ, or inf B > 3-σ. ...(*)

Suppose now that inf B > 3-σ. So σ > 3-inf B, and 3-inf B can play σ' in the definition of σ. So there exists aεA such that a > 3-inf B. So inf B > 3-a, i.e. there exists bεB such that inf B > b. This is a contradiction. So the supposition above is wrong and from (*) we conclude that inf B = 3 - sup A.

21. Notice that even if a < K for all aεA, it does not follow that sup A < K. For example, take A =]$\overline{1,3}$[and K = 3. In such a case the best conclusion possible is that sup A $\leq$ K. Such blunting of inequalities often occurs in connection with limiting processes. (Compare §54.)

FUNCTIONS

22. A _function_ has three basic features - a _domain_, a _codomain_ and a _rule_. The domain and codomain are sets and to _every_ member of the domain the rule assigns a member of the codomain. (It is possible that a particular member of the codomain may be assigned to no member of the domain, to one member of the domain or to more than one member of the domain.) A function may also be called a _mapping_.

For example, consider the function f: R $\rightarrow$ R defined by f(x) = x^2+x+1. We may also refer to f as "the function x^2+x+1" without explicit mention of the domain or codomain.

It is important to realise that a would be rule like f(x) = 1$\pm\sqrt{x}$ for all xε[0,∞[does _not_ define a function, because it attempts to associate more than one point of the codomain with certain points of the domain.

In this book we are concerned mainly with <u>real functions</u> (domain and codomain subsets of R) and <u>complex functions</u> (domain and codomain subsets of C), but keep in mind that the idea is more general.

23. Let $f: D \to E$ be a function. Let $f(D) = \{f(x): x \varepsilon D\}$. Then $f(D)$ is the <u>image</u> (or <u>range</u>) <u>of f</u>.

<u>Definition</u>. The function f is called <u>surjective</u> (or <u>onto</u>) if the image of f is its codomain, i.e. if every value in the codomain is taken by f.

For example, $f: R \to [0,\infty[$ defined by $f(x) = x^2$ is surjective because the image of f is $[0,\infty[$, whereas $g: R \to R$ defined by $g(x) = x^2$ is <u>not</u> surjective because the image of g is $[0,\infty[$ and not R.

<u>Definition</u>. The function $f: D \to E$ is <u>injective</u> (or <u>one to one</u>) if whenever $f(x_1) = f(x_2)$ it then follows that $x_1 = x_2$. (In other words, and possibly easier to grasp, f is <u>not injective</u> if there exist distinct points $x_1, x_2 \varepsilon D$ such that $f(x_1) = f(x_2)$.)

For example, $f: R \to R$ defined by $f(x) = x^2$ is <u>not</u> injective because $f(2) = f(-2)$, while $f: [0,\infty[\to R$ defined by $f(x) = x^2$ is injective.

<u>Definition</u>. A function $f: D \to E$ which is both surjective and injective is called <u>bijective</u>.

For example, $f: R \to R$ defined by $f(x) = x^3$ is bijective.

24. BOUNDEDNESS

Let f be a real function defined on D with range E.

<u>Definition</u>. f is <u>bounded above on D</u> $\iff$ E is bounded above,
i.e. $\exists M$ (a real number) s.t. $\forall x \varepsilon D$, $f(x) \le M$.
The number M is called an <u>upper bound for f</u>.

<u>Definition</u>. f is <u>bounded below on D</u> $\iff$ E is bounded below,
i.e. $\exists m$ (a real number) s.t. $\forall x \varepsilon D$, $f(x) \ge m$.
The number m is a <u>lower bound for f</u>.

<u>Definition</u>. f is bounded on D $\iff$ E is bounded,
i.e. $\exists K$ s.t. $\forall x \varepsilon D$, $|f(x)| \le K$.

(Notice that this is equivalent to $-K \leq f(x) \leq K$. However the modulus form is easier to handle in practice.)

EXAMPLE. The function x^3 is unbounded on R. The function $6x/(x^2+1)$ is bounded on R. The function $f:]0,\infty[\rightarrow R$ defined by $f(x) = 1/x^2$ is bounded below (by 0), unbounded above and unbounded.

25. SUPREMUM OF A REAL FUNCTION

Suppose that the real function f has domain D and range E, and that f is bounded above. Let sup E be denoted by σ. Then write

$$\sigma = \sup_D f \qquad \text{or just} \qquad \sigma = \sup f.$$

So then σ meets the following two requirements:

(1) $f(x) \leq \sigma \qquad \forall\, x\varepsilon D$,

(2) $\forall\, \sigma'$ less than σ, $\exists\, x\varepsilon D$ with $f(x) > \sigma'$.

If there exists $a\varepsilon D$ such that $f(a) = \sigma$, we say that f <u>attains</u> its supremum at a. A function may or may not attain its supremum.

EXAMPLE. (a) $f: R \rightarrow R$ defined by $f(x) = x^3$. Here sup f does not exist.

(b) $f: [0,1] \rightarrow R$ defined by $f(x) = 3x$. Here sup f = 3 (attained).

(c) $f:]0,1[\rightarrow R$ defined by $f(x) = 3x$. Here sup f = 3 (<u>not</u> attained).

(d) $f: R \rightarrow R$ defined by $f(0) = -2$ and $f(x) = -1/x^2$ when $x \neq 0$. Here sup f = 0 (<u>not</u> attained).

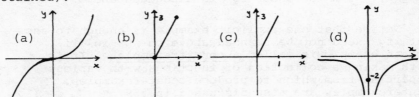

26.

Notice that if for a function f with domain D, we have that $f(x) \leq M$ for all $x\varepsilon D$, then automatically sup f $\leq$ M, because sup f is the <u>least</u> of all upper bounds for f.

Moreover if functions f, g both have domain D and $f(x) \geq g(x)$ for all $x\varepsilon D$, then sup f $\geq$ sup g. **This is because** sup f being an upper bound for f **is automatically** an upper bound for g.

Even if f(x) > g(x)
for all x∈D, it does <u>not</u> follow
that sup f > sup g.   The best
conclusion possible is that
sup f ≥ sup g.   For example,
define f, g on]0,∞[by
f(x) = 0, g(x) = -1/x.   Then
sup f = sup g = 0 even though
f(x) > g(x) for all x ε]0,∞[.

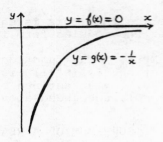

27. Notice that f+g, fg, f/g, max(f,g) and f₀g
     denote the functions whose values at x are
f(x)+g(x), f(x)g(x), f(x)/g(x), max(f(x),g(x)) and
f(g(x)) provided these are defined.   Notice
particularly the difference between fg and f₀g.

28. <u>EXAMPLE</u>.   Let f,g be functions that are bounded
     above on D.   Prove that f+g is bounded above
and that

$$\sup(f+g) \leq \sup f + \sup g.$$

Give an example to show that fg need not be bounded
above on D.

<u>Solution</u>.   (i) f(x) ≤ sup f and g(x) ≤ sup g ∀x∈D.

So f(x)+g(x) ≤ sup f + sup g ∀x∈D.   So sup f + sup g

is an upper bound for f+g.   So sup(f+g) ≤ sup f + sup g,

by §20.

     (ii) Take f(x) = -1 and g(x) = -1/x for

x∈]0,∞[.   Then f and g are bounded above by 0 but

f(x)g(x) = 1/x so that fg is unbounded above.

     Notice that one explicit example is enough to show
that fg need not be bounded above on D in (ii).
Such an example is called a <u>counterexample</u>.   It is
worth making a collection of functions with various
properties from which to produce counterexamples.
Counterexamples are also discussed in §59.

29. REAL QUADRATIC FUNCTIONS

     These are functions f: R → R given by
$f(x) = ax^2+bx+c$, where a,b,c∈R and a ≠ 0.
     The graph of such a function f is a parabola with
axis parallel to the y-axis.   The parabola has <u>hanging</u>
shape or <u>inverted</u> shape according as a > 0 or a < 0.
This division is a consequence of the dominance of the
ax^2 term for large values of x.   (See §12.)
The possible shapes are shown on page 13.

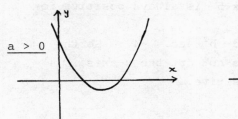
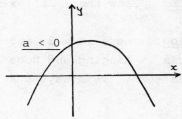

Also important is whether the parabola cuts the
x-axis. The number b^2-4ac settles this as the
following result shows.

RESULT.   The graph of the quadratic function defined
above
        (i)   cuts the x-axis in two distinct points
if b^2-4ac > 0,
        (ii)   touches the x-axis at one point if
b^2-4ac = 0,
        (iii)   fails to meet the x-axis if b^2-4ac < 0.

Proof.   Notice that

$$f(x) = ax^2 + bx + c = a\left(x + \frac{b}{2a}\right)^2 + \left(c - \frac{b^2}{4a}\right)$$

$$= a\left(x + \frac{b}{2a}\right)^2 - \frac{1}{4a}(b^2 - 4ac).$$

At a point x for which f(x) = 0, we therefore have

$$4a^2\left(x + \frac{b}{2a}\right)^2 = b^2 - 4ac. \qquad \ldots(*)$$

The left-hand side of (*) being a perfect square is
non-negative. So (*) is impossible if $(b^2 - 4ac) < 0$.
This proves case (iii).

    If $(b^2 - 4ac) = 0$, there is only one value of x
satisfying (*), namely $x = -\frac{b}{2a}$.   This proves case (ii).

    If $(b^2 - 4ac) > 0$, there are two solutions of (*),
which is case (i).

N.B.   The maximum or minimum value of the quadratic
        function f is attained when $x = -\frac{b}{2a}$.

EXAMPLE. Show that $2x^2-6x+5$ is always positive for $x\epsilon R$.

Solution. Here $a = 2$ and $b^2-4ac = -4$. So the parabola is hanging and does not cut the x-axis, i.e. $2x^2-6x+5$ is always positive.

EXAMPLE. Show that $-x^2+x-1$ and $5x^4-3x^2+1$ are respectively always negative and positive for $x\epsilon R$.

EXAMPLE. Show that if $a,b,c\epsilon R$ and a and c have opposite signs, then the equation $ax^2+bx+c = 0$ has two distinct real roots.

30. INVERSE FUNCTIONS

Corresponding to a __bijective__ function $f: X \to Y$, there is a function $f^{-1}: Y \to X$, called the __inverse of__ f and defined by $f^{-1}(y) = x$ if and only if $f(x) = y$.

The function f^{-1} is unique and f^{-1} can only be defined if f is bijective. Notice that f^{-1} reverses the action of f in the sense that $f^{-1} \circ f = i$, where i denotes the identity function on X defined by $i(x) = x$ for all $x\epsilon X$.

EXAMPLE. Let X, Y, Z be any sets. Let $g: X \to Y$ and $f: Y \to Z$ be bijective functions. Prove that (i) $f \circ g: X \to Z$ is bijective, and (ii) $(f \circ g)^{-1} = g^{-1} \circ f^{-1}$.

Solution. (i) (__To show__ $f \circ g$ __surjective__) Let $z\epsilon Z$. Then since f is surjective, $z = f(y)$ for some $y\epsilon Y$. But then, since g is surjective, $y = g(x)$ for some $x\epsilon X$. So then $(f \circ g)(x) = f(g(x)) = f(y) = z$. So $f \circ g$ is surjective.

(__To show__ $f \circ g$ __injective__) Suppose that $x_1, x_2 \epsilon X$ and $(f \circ g)(x_1) = (f \circ g)(x_2)$. So $f(g(x_1)) = f(g(x_2))$. Since f is injective it then follows that $g(x_1) = g(x_2)$. But then since g is injective it follows that $x_1 = x_2$. So we conclude that $f \circ g$ is injective.

It follows that $f_{\mathbf{o}}g$ is bijective and that $(f_{\mathbf{o}}g)^{-1}$ exists.

(ii)   Clearly $(f_{\mathbf{o}}g)^{-1}$ and $g^{-1}{}_{\mathbf{o}}f^{-1}$ both map Z into X.   Take an arbitrary element $z\varepsilon Z$.   From the bijective character of f and g, there exists $y\varepsilon Y$ such that $f(y) = z$ and in turn there exists $x\varepsilon X$ such that $g(x) = y$.   Then

$$(f_{\mathbf{o}}g)(x) = z \text{ so that } (f_{\mathbf{o}}g)^{-1}(z) = x.$$

On the other hand

$$(g^{-1}{}_{\mathbf{o}}f^{-1})(z) = g^{-1}(f^{-1}(z)) = g^{-1}(y) = x.$$

So $(f_{\mathbf{o}}g)^{-1}$ agrees with $g^{-1}{}_{\mathbf{o}}f^{-1}$ everywhere on Z.
So $(f_{\mathbf{o}}g)^{-1} = g^{-1}{}_{\mathbf{o}}f^{-1}$.

QUANTIFIERS AND LOGIC

31.   We use the symbol $\forall$ with the meaning "<u>for all</u>" or "<u>for every</u>", and we use the symbol $\exists$ with the meaning "<u>there exists a</u>" or "<u>there exists at least one</u>". These are the <u>universal and existential quantifiers</u> respectively.   We also use the symbol $\neg$ to mean "<u>not</u>". Notice that translating $\forall$ as "<u>for any</u>" can lead to trouble because <u>any</u> does not always mean <u>every</u>. Compare:

In <u>any</u> city you will find a library.   (any = every)

Find <u>any</u> piece of wreckage that will float.
(any $\neq$ every)

32.   NEGATION OF QUANTIFIERS

(i)   To negate a statement with quantifiers, change $\forall$ to $\exists$, change $\exists$ to $\forall$ and negate the main verb.   As evidence of this consider the following examples:

(A)   Not all ships have a metal deck.

i.e. $\neg(\forall$ships, the deck is metal$)$.

i.e. $\exists$ a ship such that the deck is <u>not</u> metal.

(B)   There do not exist ships such that the deck
is cardboard.

i.e. $\neg(\exists$ships such that the deck is cardboard$)$.

i.e. $\forall$ ships the deck is <u>not</u> cardboard.

From (A) and (B) we can see that

(1) $\neg(\forall x, \ P(x))$ is $\exists x$ such that $\neg P(x)$,

and (2) $\neg(\exists x$ such that $P(x))$ is $\forall x, \ \neg P(x)$.

EXAMPLE.   <u>f</u> <u>is bounded above</u> means:

$\exists$ M   s.t. $\forall$xϵD,   f(x) $\leq$ M.

<u>f</u> <u>is not bounded above</u> means:

$\forall$M, $\exists$xϵD s.t. f(x) > M.

(ii) Notice also that

$\neg$(A <u>and</u> B) is equivalent to ($\neg$ A <u>or</u> $\neg$B),

and $\neg$(A <u>or</u> B) is equivalent to ($\neg$ A <u>and</u> $\neg$ B).

These facts are just the De Morgan Laws of set theory:

(C $\cap$ D)' = C' $\cup$ D' and (C $\cup$ D)' = C' $\cap$ D'.

For example,

f is unbounded $\Longleftrightarrow$ $\neg$(f is bounded)

$\Longleftrightarrow$ $\neg$(f is bounded above <u>and</u> f is bounded below)

$\Longleftrightarrow$ $\neg$(f is bounded above) <u>or</u> $\neg$(f is bounded below)

$\Longleftrightarrow$ f is unbounded above <u>or</u> f is unbounded below.

33. ORDER OF QUANTIFIERS

In a sentence where both $\exists$ and $\forall$ appear (as in the definition of a function being bounded above), the <u>order</u> of the quantifiers cannot in general be altered.   <u>Notice</u> the difference in meaning between

(1) $\forall$ ships $\exists$ a colour s.t. the funnel is of that
colour,

and (2) $\exists$ a colour s.t. $\forall$ ships the funnel is of that
colour.

In the real world·(1) is true but (2) is false.

Notice however the equivalence of

$\exists$ M s.t. $\forall$xϵD, f(x) $\leq$ M <u>and</u> $\exists$M s.t. f(x) $\leq$ M $\forall$xϵD.

34. NECESSARY AND SUFFICIENT CONDITIONS

Consider the following statements about a positive integer N.

A. N has 0 as its last digit.

B. N is divisible by 5.

C. N is divisible by 100.

D. N is even and is divisible by 5.

Now look at what follows.

(1)   B is a necessary condition for A, but it is not a sufficient condition for A,

i.e.   B $\Leftarrow$ A   but   B $\not\Rightarrow$ A,

i.e.   B is implied by A, but B does not imply A.

To see that B $\not\Rightarrow$ A, take N = 15 so that B is true while A is false.

(2)   C is a sufficient condition for A, but it is not a necessary condition for A,

i.e.   C $\Rightarrow$ A   but   C $\not\Leftarrow$ A,

i.e.   C implies A, but C is not implied by A.

To see that C $\not\Leftarrow$ A, take N = 80 so that C is false while A is true.

(3)   D is both necessary and sufficient for A,

i.e.   D $\Rightarrow$ A   and   D $\Leftarrow$ A,

i.e.   D and A imply each other.

We can then say that D is true <u>if and only if</u> A is true and we can write D $\Leftrightarrow$ A to denote this.

The sign $\Leftrightarrow$ is useful for linking successive lines of a proof (because it has the force of a conjunction). However attempts to use $\Rightarrow$ in a similar way are generally less successful (because $\Rightarrow$ has the force of a verb). Notice <u>so</u>, <u>then</u>, <u>i.e.</u>, <u>therefore</u> as useful linking words.

35. CONTRADICTION ARGUMENTS

These may succeed if assuming the opposite of what you need to prove allows you to write down a statement or definition from which you can work.   A case in point is the proof that $\sqrt{2}$ is irrational.   Here assume that $\sqrt{2}$ is rational and you then can write down that $\sqrt{2}$ = m/n where m,nϵN.   See the solution of Ex.1.32.

The presence of a <u>negative</u> aspect in the question often sets up a suitable situation.   See also §86.

36.   <u>EXAMPLE</u>.   A is a bounded subset of R and |a-b| < 1 <u>for all</u> a,bϵA.   Prove that sup A - inf A $\leq$ 1.

<u>Solution</u>. Suppose not.   So sup A - inf A > 1 and sup A > 1 + inf A.   So 1 + inf A can play σ' in the definition of sup A.   It follows that there exists aϵA with a > 1 + inf A.   Then a-1 > inf A and a-1 can play τ' in the definition of inf A.   So there exists bϵA with a-1 > b.   So a-b > 1.   This contradicts the given hypothesis so that our assumption was wrong.   So we conclude that sup A - inf A $\leq$ 1.

37. __EXAMPLE__. Show from the definition that f defined on $R - \{0\}$ by $f(x) = 1/x^2$ is unbounded above.

__Solution__. We show that

$$\neg(\,\exists\, M \text{ such that } \forall x \in R - \{0\}, \quad f(x) \leq M),$$

i.e. $\forall M \quad \exists x \in R - \{0\}$ such that $f(x) > M.$

Assume without loss of generality that $M > 0$. (See the remark below.) Then we want $1/x^2 > M$, i.e. $x^2 < 1/M$. So, take $x = 1/(2\sqrt{M})$. Then we see that

$f(1/(2\sqrt{M})) = 4M > M.$ So f is unbounded above.

__Remark__. Assuming that $M > 0$ makes the inequalities easier to handle. No loss is involved in this assumption because if we check that f can exceed every __positive__ number M then it certainly also exceeds every negative number M.

38. __EXAMPLE__. Show that f defined on $R - \{2,9\}$ by

$$f(x) = \frac{x-4}{(x-2)(x-9)} \quad \text{is unbounded above.}$$

__Solution__. (Compare §37.) We show that

$$\forall M > 0, \quad \exists x \in D \text{ such that } f(x) > M.$$

Clearly f is large as x approaches 9 from the right. So, we seek a suitable $x \in\,]9,10[$. Then,

$$\forall x \in\,]9,10[, \quad f(x) = \frac{x-4}{(x-2)(x-9)} > \frac{5}{8(x-9)}. \quad \ldots(*)$$

We now want $x \in\,]9,10[$ with $\dfrac{5}{8(x-9)} > M$,

i.e. $9 + \dfrac{5}{8M} > x > 9$.

So take $x = 9 + \dfrac{1}{8M}$, with the further restriction

(without loss) that $M > \dfrac{1}{8}$ to keep $x \in\,]9,10[$. Then,

$$f(9 + \frac{1}{8M}) > \frac{5}{8 \cdot \frac{1}{8M}} = 5M > M.$$

So f is unbounded above.

__Notice__: In the above example we do not attempt to solve the inequality $\dfrac{x-4}{(x-2)(x-9)} > M$ as this is hard.

Instead, we deal with the easier inequality $\dfrac{5}{8(x-9)} > M$ by first making the estimate $(*)$, which retains only the vital factor $(x-9)$.

39. THE INEQUALITY OF THE ARITHMETIC AND GEOMETRIC MEANS

The <u>arithmetic mean</u> of two positive numbers a, b is $\frac{1}{2}(a+b)$. Their <u>geometric mean</u> is $\sqrt{(ab)}$. The following result is sometimes useful.

RESULT. Let a, b be positive numbers. Then their geometric mean cannot exceed their arithmetic mean, i.e.

$$\sqrt{(ab)} \le \frac{1}{2}(a+b).$$

Proof. Since the square of a real number is non-negative, $(\sqrt{a} - \sqrt{b})^2 \ge 0$, i.e. $a + b - 2\sqrt{(ab)} \ge 0$. Transpose to see that $\sqrt{(ab)} \le \frac{1}{2}(a+b)$, as required.

EXAMPLE. Show that the sum of a positive number and its reciprocal is greater than or equal to 2.

Solution. Let $a > 0$. Then, by the above result,

$$\sqrt{(a \cdot \frac{1}{a})} \le \frac{1}{2}\left(a + \frac{1}{a}\right).$$

So $2 \le \left(a + \frac{1}{a}\right)$ as required.

The result above can be extended to the case of n positive numbers, say, $a_1, a_2, \ldots, a_n$. It is then true that

$$(a_1 a_2 \ldots a_n)^{1/n} \le \frac{a_1 + a_2 + \ldots + a_n}{n}$$

and further, that if $k_1, k_2, \ldots, k_n$ are positive numbers such that $k_1 + k_2 + \ldots + k_n = 1$ then

$$a_1^{k_1} a_2^{k_2} \ldots a_n^{k_n} \le k_1 a_1 + k_2 a_2 + \ldots + k_n a_n.$$

For a proof of these results see the books by Rankin and by Rudin mentioned on page 221.

EXAMPLE. Let a, b, c be positive numbers. Prove that $a^3 + b^3 + c^3 \ge 3abc$ and that $\frac{a}{b} + \frac{b}{c} + \frac{c}{a} \ge 3$.

EXAMPLES 1

(All numbers in these examples are real.)

1. Let $a > b$ and $c < 0$. Prove from the conditions of §2 that $ca < cb$.

2. If $0 < a < b$, prove that $\dfrac{a}{y+a} < \dfrac{b}{y+b}$ for all $y > 0$.

3. Let $|x| \le a$ and $|y| \le a$. Prove that
    (i) $|x^2 - y^2| \le 2a|x - y|$,
    (ii) $|x^3 - y^3| \le 3a^2|x - y|$,
    (iii) $|x^n - y^n| \le na^{n-1}|x - y|$, for all $n \epsilon N$.

4. If $|x - a| < \frac{1}{2}|a|$, show that $\frac{1}{2}|a| < |x| < \frac{3}{2}|a|$.

5. Use some form of the triangle inequality to show that if $|x| \le 1$ then $|x^6 - 4x - 17| \ge 12$.

6. Show that $|a + b| \ge ||a| - |b||$.

7. The real number a is such that $ab < 2$ for all b with $0 < b < 1$. Prove by a contradiction argument that $a \le 2$.

8. Show that
    $$\left|\frac{(x-3)}{(x-2)(x-8)}\right| \ge \frac{4}{7|x-8|} \quad \text{for} \quad x \epsilon [7,9] - \{8\}.$$

9. Show that $\left|\dfrac{x}{1-x}\right| \le 4|x|$ when $|x| \le \dfrac{3}{4}$.

10. Prove by estimating that $180x/(x^2+7)^2$ exceeds 1 for all $x \epsilon [3,4]$.

11. Show that $\left|\dfrac{2}{n^3} + \dfrac{5(-1)^n}{n^2} + \dfrac{1}{n}\right| \le \dfrac{8}{n}$ for all $n \epsilon N$.

12. In each of the following cases, find a positive constant K such that an estimate of the type given holds for all positive integers n greater than some number X. (N.B. It is K not X you are to find.)

(a) $\dfrac{3n+1}{n^2-5n+1} \le \dfrac{K}{n} \qquad \forall n > X.$

(b) $\dfrac{2n^2-13n}{n+3} \ge Kn \qquad \forall n > X.$

(c) $\left|\dfrac{6n^2-n+1}{2n^2-n+5} - 3\right| \le \dfrac{K}{n} \qquad \forall n > X.$

13. Prove that there exists a real number X such that $\frac{2}{n} \leq \frac{3n+2}{n^2-3n+1} \leq \frac{10}{n}$ for all positive integers n greater than X.

14. Prove by induction that

$$1 + 2 + 3 + \ldots + n = \tfrac{1}{2}n(n+1),$$

$$1^2 + 2^2 + 3^2 + \ldots + n^2 = \tfrac{1}{6}n(n+1)(2n+1),$$

$$1^3 + 2^3 + 3^3 + \ldots + n^3 = \left(\tfrac{1}{2}n(n+1)\right)^2.$$

15. A is a bounded subset of R. B is the set defined by B = {b: b = 2a+3, aϵA}. Prove that sup B = 2 sup A + 3.

16. A bounded subset A of R has the property that $|a-b| < 1$ for every pair of elements a,bϵA. Prove by a contradiction argument or otherwise that (sup A - inf A) $\leq$ 1.

17. A is a bounded non-empty set of positive numbers.   Let B = {b: b = 1/a, aϵA}.   Show that B is bounded below and that inf B = 1/sup A.
   If inf A = 0, show by a contradiction argument that B is not bounded above.

18. Give an example of bounded sets A, B $\subseteq$ R, which do not intersect, which do not contain their suprema, but for which sup A = sup B.

19. Let f: R $\to$ R and g: R $\to$ R be functions.
   (a) Show that if f and g are bounded then so also is fg.
   (b) Show that if g(x) > α for all xϵR, where α is a fixed <u>positive</u> number, then 1/g is bounded.
   (c) Give an example to show that if we relax the condition in (b) and demand only that g(x) > 0 for all xϵR, then 1/g need not be bounded.
   (d) Show that if f is bounded then f $\circ$ g is also bounded.

20. Prove that f defined on R - {3} by

$$f(x) = \frac{x^2+5}{(x^2+2)(x-3)}$$

is unbounded above.

21.   Prove that f defined on R - {4,6} by

$$f(x) = \frac{x^2+2}{(x-4)(x-6)^2}$$

is unbounded.

22.   Give an example to show that if f and g are unbounded functions on]0,∞[, then it is possible for fg to be bounded.

23.   Show that if f and g are bounded then so also are f+g and f-g.   Use a contradiction argument to prove that if h is unbounded and k is bounded, then (h+k) is unbounded.

24.   Prove that if f is unbounded and g is bounded with inf g > 0, then fg is unbounded.

25.   Let f: R → R be a bounded function.   Let A ⊆ B ⊆ R.   Prove that

$$\sup_{A} f \le \sup_{B} f \quad \text{and} \quad \inf_{A} f \ge \inf_{B} f.$$

26.   Let f and g be real functions that are bounded above on a domain D.   Show that (f+g) is bounded above on D and that

$$\sup (f+g) \le \sup f + \sup g.$$

Give an example to show that equality need not hold. Give also an example to show that f^2 need not be bounded above.

27.   Let f and g be bounded positive functions on a domain D.   By considering the relation f(x)g(x) ≤ sup (fg), prove that g(x) inf f ≤ sup (fg) for all x∈D.   Deduce that inf f.sup g ≤ sup (fg).   Deduce further that $\inf (fg) \le (\sup f.\inf f)^{\frac{1}{2}} (\sup g.\inf g)^{\frac{1}{2}}$ ≤ sup (fg).

28.   Give an example of a bounded non-constant rational function.

29.   The functions f: R → R and g: R → R are bounded.   Show that (sup f - inf g) is an upper bound for (f-g) and deduce that sup (f-g) ≤ sup f - inf g.
      By considering the relation

$$f(x)-g(x) \le \sup (f-g) \quad \text{for all} \quad x∈R,$$

prove that

$$\sup f - \sup g \le \sup (f-g).$$

30.   Use the inequality of the means to show that for all positive numbers x and α,

$$\frac{x^2+\alpha}{2x} \geq \sqrt{\alpha}.$$

31.   Use the inequality of the means to show that for every integer r with $1 \leq r \leq 2n$, $\sqrt{(r(2n+1-r))} \leq (n+\tfrac{1}{2})$.

Deduce that $\displaystyle\sum_{r=1}^{2n} \frac{1}{\sqrt{(r(2n+1-r))}} \geq \frac{4n}{2n+1}$ for every $n \varepsilon N$.

32.   Prove that $\sqrt{2}$ is irrational.

33.   Show that in the field Z_5 there is no ordering > satisfying the conditions 1,2,3 of §2.   (See §2 for the definition of Z_5.)

II.   LIMITS OF SEQUENCES

---

40.   A <u>sequence</u> is a function defined on N   (or on a subset of N).

For example, take f: N → R, defined by
f(n) = (5n+4)/n.   This sequence can be denoted by f, by {f(n)}, by {(5n+4)/n} or by exhibiting a few terms as in {9,7,19/3,6,29/5,...}.

41.   THE LIMIT OF A SEQUENCE AS N TENDS TO INFINITY

<u>Definition</u>.   Let {f(n)} be a sequence.   Let A be a <u>fixed real</u> number.   Then {f(n)} <u>tends to the limit A as n tends to infinity</u> means:

$$\forall \varepsilon > 0, \quad \exists X \text{ s.t. } \quad |f(n)-A| < \varepsilon \quad \forall n > X.$$

[<u>Write</u>: f(n) → A as n → ∞, or $\lim_{n \to \infty} f(n) = A$.]

Notice: The definition says that for every positive number ε the values f(n) must lie in the band]A − ε, A + ε[for all n beyond some threshold X.   In the figure opposite the points are the values f(n).   Notice that beyond X all these points lie in the band from A − ε to A + ε.

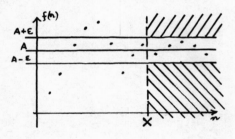

There is no necessity for X to be the least positive number beyond which |f(n)−A| < ε holds.
Think of ε as a small positive number which is chosen first and then dictates the value of X.   So in general X will depend on ε and the smaller ε is taken, the larger X will have to be.   Notice that this happens in the example of §42(i).

42.   (i)   <u>EXAMPLE</u>.   Let f be defined on N by
$$f(n) = 5 + \frac{4}{n}.$$ Prove that f(n) → 5 as n → ∞.

<u>Solution</u>.   Choose ε > 0 and take A = 5 in the definition.   Then,

$$|f(n)-5| = |5 + \frac{4}{n} - 5| = \frac{4}{n} < \varepsilon \quad \forall n > \frac{4}{\varepsilon}.$$

(Notice that in this case A = 5 and X = 4/ε.   Notice also that as ε is made smaller so the value of X increases.)

(ii) EXAMPLE.   Let f be defined as in (i). Prove that $f(n) \not\to 3$ as n → ∞.

Solution.   Suppose that f(n) → 3 as n → ∞.   Then take ε = 1 in the definition to conclude that

$$\exists X \quad \text{such that} \quad |f(n)-3| < 1 \quad \forall n > X.$$

$$\text{So} \quad 2 + \frac{4}{n} < 1 \quad \forall n > X,$$

$$\text{i.e.} \quad \frac{4}{n} < -1 \quad \forall n > X.$$

This is impossible.   So the original assumption is false.   So $f(n) \not\to 3$ as n → ∞.

43. It is important to appreciate the force of the phrase $\forall \varepsilon > 0$ in the definition of the limit. In §42(i), i.e. when A = 5, no matter what value of ε > 0 is chosen, we can produce a threshold X (namely 4/ε).   This corresponds to being able to choose as narrow a band as we like containing 5 and still know that beyond some point all values f(n) lie in this band. In contrast, in §42(ii), i.e. when A = 3, though it is true that for certain ε > 0 (namely all ε > 2) we can produce such a threshold (namely 4/(ε-2)), there are many values of ε > 0 (namely all ε with 0 < ε ≤ 2) for which no such threshold can be found.   This means that for certain bands containing 3 it is impossible to find a threshold beyond which all the values f(n) lie in the band.   This disqualifies 3 as a possible limit of {f(n)}.

44. The value of a limit can often be seen from an examination of the dominant components of the nth term.

EXAMPLE.   Examine the behaviour as n → ∞ of {f(n)} where $f(n) = \frac{4n^2 + 9}{5n^2 - 7n + 11}$.

B

<u>Solution</u>.   (The dominant terms on the top and bottom are $4n^2$ and $5n^2$ respectively.   So the limit is likely to be $4/5$.)

Let $\varepsilon > 0$.   Then

$$|f(n) - \tfrac{4}{5}| = \left| \frac{28n + 1}{5(5n^2 - 7n + 11)} \right|$$

$$\leq \frac{29n}{5 \cdot \tfrac{5}{2} n^2} \qquad \forall n > Y \quad \text{(by §12)}$$

$$< \varepsilon \qquad \forall n > \max(Y, \tfrac{58}{25\varepsilon}).$$

So $f(n) \to 4/5$ as $n \to \infty$.

<u>EXAMPLE</u>.   Examine the behaviour as $n \to \infty$ of

(a) $\dfrac{6n^2 - 7n}{3n^2 - 14n - 2}$,   (b) $\dfrac{n^2 + 1}{4n^3 - 31n}$.

We can say that "$3n^2 - 7n + 1$ <u>behaves</u> like $3n^2$ as $n \to \infty$".   If you want a notation to express this use the $\sim$ notation of §267 in the statement "$3n^2 - 7n + 1 \sim 3n^2$ as $n \to \infty$".   Do not however make statements like "$3n^2 - 7n + 1 \to 3n^2$ as $n \to \infty$": these do not fall within the scope of the definition of the $\to$ sign, because the right-hand side depends on n.

45.   SEQUENCES WHICH TEND TO INFINITY AS N TENDS TO INFINITY

<u>Definition</u>.   Let $\{f(n)\}$ be a sequence.   Then $\{f(n)\}$ <u>tends to infinity as n tends to infinity</u> means:

$$\forall K \varepsilon R, \ \exists X \text{ s.t. } f(n) > K \quad \forall n > X.$$

[<u>Write</u>: $f(n) \to \infty$ as $n \to \infty$.]

Notice: The definition says that for every fence height K, there is a threshold X such that f(n) lies above K for all values of n beyond X.   For example if $f(n) = 2n^2 + 9$, then $f(n) \to \infty$ as $n \to \infty$.
    If it suits us, we can without loss of generality look only at all K > 0 or at all K > 1, as functions which can jump high fences can also jump low ones.   In general, X is dictated by K and X tends to increase as K increases.   See the example in §46.
    The phrase, $\forall K \varepsilon R$, gives force to the definition. It is this phrase that allows us to keep raising the fence height K and so force the function higher and higher.

In the diagram the points are the values f(n). Notice that for all n beyond X the points f(n) lie above the value K.

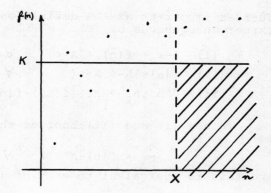

There is a similar definition to the one above for sequences that tend to minus infinity.

46. <u>EXAMPLE</u>. Examine the behaviour as $n \to \infty$ of $\{f(n)\}$ where $f(n) = \dfrac{n^2-5n+9}{3n+13}$.

<u>Solution</u>. (The dominant terms on the top and bottom are n^2 and $3n$ respectively. So f(n) behaves like n/3 as $n \to \infty$. So $\{f(n)\}$ is likely to tend to infinity as $n \to \infty$.)

Without loss of generality take K > 0. (Cf. §37.)

Then $f(n) = \dfrac{n^2-5n+9}{3n+13} \geq \dfrac{\frac{1}{2}n^2}{3n+13n} \quad \forall n > Y$

$$= \frac{n}{32} > K \quad \forall n > \max(Y, 32K).$$

So $\forall K > 0$, $f(n) > K \quad \forall n > \max(Y, 32K)$.

So $f(n) \to \infty$ as $n \to \infty$.

N.B. Taking K > 0 in examples of this type often facilitates handling of inequalities.

<u>EXAMPLE</u>. Examine the behaviour as $n \to \infty$ of $\{f(n)\}$ where $f(n) = \dfrac{n^3-12n-2}{5n+11}$.

47. Use the word <u>infinity</u> only in the phrase <u>tends to infinity</u>. Attempts to treat infinity as a real number as in $1/\infty = 0$ often lead to trouble.

48. OTHER ASPECTS OF THE DEFINITION OF A LIMIT

Suppose $f(n) \to A$ as $n \to \infty$. Then in the usual way
$$|f(n)-A| < \varepsilon \quad \forall n > X. \qquad \ldots(*)$$

Just as important as the definition are the following three consequences of (*):

(i) $A-\varepsilon < f(n) < A+\varepsilon$ $\forall\, n > X$.

(ii) $f(n)\varepsilon\,]A-\varepsilon, A+\varepsilon\,[$ $\forall\, n > X$.

(iii) $\forall\, n_1, n_2 > X$, $|f(n_1)-f(n_2)| < 2\varepsilon$.

To see (i) and (ii), notice that (*) is equivalent to

$$-\varepsilon < f(n)-A < \varepsilon \qquad \forall\, n > X$$

and this is equivalent to each of (i) and (ii).

To see (iii), notice that by triangle inequality

$$|f(n_1)-f(n_2)| \le |f(n_1)-A| + |A-f(n_2)|$$
$$< 2\varepsilon \qquad \forall\, n_1, n_2 > X.$$

In many situations (i), (ii) and (iii) can be more useful than the basic definition of §41.

49. BOUNDEDNESS

A sequence is called <u>convergent</u> if it tends to a <u>finite</u> limit.   The following result is important.

<u>RESULT</u>.   A convergent sequence is bounded.

<u>Proof</u>.   Suppose $f(n) \to A$ as $n \to \infty$ and take $\varepsilon = 1$ in the definition to deduce from §48(ii) that $f(n)\varepsilon\,]A-1, A+1\,[$ for all $n \ge P$, where we can assume without loss of generality that $P\varepsilon N$.   Then an upper bound for $\{f(n)\}$ is $\max(f(1), f(2), \ldots, f(P), A+1)$ and a lower bound is $\min(f(1), f(2), \ldots, f(P), A-1)$.   So $\{f(n)\}$ is bounded.

Notice that real functions and sequences differ in this respect.   A real function like $f(x) = (3x+1)/x$ for $x\varepsilon\,]0, \infty\,[$ tends to the limit 3 as $x \to \infty$, but f <u>is</u> unbounded, the unbounded behaviour occurring close to $x = 0$.   A sequence (like $g(n) = (3n+1)/n$) is defined only on N <u>not</u> on $]0, \infty\,[$ and so is denied the possibility of unbounded behaviour close to $x = 0$.

50. THE USE OF MULTIPLES OF EPSILON

Suppose that $f(n) \to A$ as $n \to \infty$. Then if asked about the behaviour of the sequence $\{7f(n)\}$ as $n \to \infty$, we can proceed thus:

$\forall \epsilon > 0,$ $\exists X$ such that $|f(n) - A| < \epsilon$ $\forall n > X.$

So, $\forall \epsilon > 0,$ $\exists X$ such that $|7f(n) - 7A| < 7\epsilon$ $\forall n > X.$

Despite the 7ϵ, the latter statement is acceptable as a proof that $7f(n) \to 7A$ as $n \to \infty$, because it holds for every $\epsilon > 0$, and we can still cut down the gap between $7f(n)$ and $7A$ as small as we like by choosing ϵ to be sufficiently small. So, in general, the statement

$\forall \epsilon > 0,$ $\exists X$ such that $|g(n) - B| < M\epsilon$ $\forall n > X$

is sufficient to prove that $g(n) \to B$ as $n \to \infty$ provided that M is a genuine positive constant, (i.e. provided that M is positive and is independent of n.)

51. HEREDITY RESULTS FOR LIMITS

RESULT. Suppose that $f(n) \to A$ and $g(n) \to B$ as $n \to \infty$, and that k is a real number. Then, as $n \to \infty$,

$kf(n) \to kA,$ $|f(n)| \to |A|,$ $f(n)+g(n) \to A+B,$ $f(n)g(n) \to AB.$

Also $f(n)/g(n) \to A/B$ provided that $B \neq 0$.

Proof. Choose $\epsilon > 0$. Then there exist X_1, X_2 with

$|f(n) - A| < \epsilon$ $\forall n > X_1$ and $|g(n) - B| < \epsilon$ $\forall n > X_2.$

So $|kf(n) - kA| < |k|\epsilon$ $\forall n > X_1$, i.e. the first result. Also $||f(n)| - |A|| \leq |f(n) - A| < \epsilon$ $\forall n > X_1$, i.e. the second result. Then, for the sum notice that

$|(f(n) + g(n)) - (A + B)| \leq |f(n) - A| + |g(n) - B| < 2\epsilon$
$$\forall n > \max(X_1, X_2).$$

For the product notice that
$|f(n)g(n) - AB| = |f(n)g(n) - Ag(n) + Ag(n) - AB|$
$\leq |g(n)||f(n) - A| + |A||g(n) - B| \leq M\epsilon + |A|\epsilon$ $\forall n > \max(X_1, X_2),$
where M is an upper bound for the convergent sequence $\{|g(n)|\}$ by §49.

For the quotient it is enough to show that $1/g(n) \to 1/B$ as $n \to \infty$, because $f(n)/g(n)$ can be taken as the product of $f(n)$ and $1/g(n)$. Notice first from above that $|g(n)| \to |B|$, which we are assuming non-zero. So use §48(i) with $\epsilon = \frac{1}{2}|B|$ to get $|g(n)| > \frac{1}{2}|B|$ $\forall n > X_3.$

So $\quad \left| \dfrac{1}{g(n)} - \dfrac{1}{B} \right| = \dfrac{|g(n) - B|}{|g(n)||B|} < \dfrac{\varepsilon}{\frac{1}{2}|B|^2} \qquad \forall n > \max(X_2, X_3).$

So $\quad 1/g(n) \to 1/B \quad$ as $\quad n \to \infty \quad$ as we wished to show.

The results for $f - g$, f^2, f^3 and so on follow from the above result. Certain analogues of these results also exist for sequences that tend to infinity. For example, if $g(n) \to \infty$ as $n \to \infty$, then $1/g(n) \to 0$.

52. The result of §51 shows that if $f(n) \to A$ as $n \to \infty$, then $(f(n))^2 \to A^2$, and by extension, for every fixed positive integer k, $(f(n))^k \to A^k$. However, statements like $(f(n))^n \to A^n$, where the right-hand side depends on n are invalid and can reach wrong conclusions. For example, $\left(1 + \dfrac{1}{n}\right) \to 1$ as $n \to \infty$ but we see in §214 that $\left(1 + \dfrac{1}{n}\right)^n \to e = 2.718\ldots$, as $n \to \infty$.

53. Notice how the heredity results of §51 can be used to avoid an ε-type argument, as in the following.

<u>EXAMPLE</u>. Find $\lim\limits_{n\to\infty} \dfrac{3n^2 - 7n + 6}{2n^2 + 4}$.

<u>Solution</u>. Divide above and below by n^2. So,

$$\frac{3n^2 - 7n + 6}{2n^2 + 4} = \frac{3 - 7/n + 6/n^2}{2 + 4/n^2} \to \frac{3}{2} \quad \text{as} \quad n \to \infty,$$

by the heredity result for quotients (§51).

54. SANDWICH RESULTS

(i) <u>RESULT</u>. Suppose $f(n) \le g(n) \le h(n) \quad \forall n > X$, and that $f(n) \to A$ and that $h(n) \to A$ as $n \to \infty$. Then $g(n) \to A$ as $n \to \infty$.

<u>Proof</u>. For every $\varepsilon > 0$, there exists a threshold Y such that

$$A - \varepsilon < f(n) \le g(n) \le h(n) < A + \varepsilon \qquad \forall n > Y.$$

So $g(n) \in \,]A - \varepsilon, A + \varepsilon[\quad \forall n > Y$, i.e. $g(n) \to A$ as $n \to \infty$.

(ii) <u>RESULT</u>. Suppose $f(n) \geq 0$ $\forall n > X$, and that
$f(n) \to A$ as $n \to \infty$.   Then $A \geq 0$.

<u>Proof</u>.   Suppose not and that $A < 0$.   So by §48(ii)
deduce that $f(n) \varepsilon]2A,0[$ for all $n > Y$.   So $f(n) < 0$
for all $n > Y$.   This contradicts the hypothesis that
$f(n) \geq 0$.   So $A \geq 0$.

(iii) <u>RESULT</u>.   Suppose that $f(n) \geq g(n)$ $\forall n > X$,
and that $f(n) \to A$ and $g(n) \to B$ as $n \to \infty$.   Then
$A \geq B$.

<u>Proof</u>.   Apply (ii) to $h(n)$ where $h(n) = f(n) - g(n)$.

In connection with (iii), notice that even if
$f(n) > g(n)$ for all $n \geq 1$ there is no guarantee that
$A > B$.   For example, if $f(n) = 1/n$ and $g(n) = 0$ for
all $n \varepsilon N$, then $f(n) > g(n)$ for all $n \geq 1$ but
$A = B = 0$.
There are other analogues of these results.   For
example if $f(n) \geq g(n)$ for all $n > X$ and $g(n) \to \infty$
as $n \to \infty$, then $f(n) \to \infty$ as $n \to \infty$.

<u>EXAMPLE</u>.   Define $f: N \to R$ by
$$f(n) = \frac{1}{n^2+n+1} + \frac{1}{n^2+n+2} + \ldots + \frac{1}{n^2+3n+2}.$$
Discuss the behaviour of $\{f(n)\}$ as $n \to \infty$.

<u>Solution</u>.   The expression for $f(n)$ contains $(2n+2)$
terms, each at least $1/(n^2+3n+2)$ and at most $1/(n^2+n+1)$.
So
$$\frac{2n+2}{n^2+3n+2} \leq f(n) \leq \frac{2n+2}{n^2+n+1} \text{ for all } n \varepsilon N.$$
The LHS and RHS both tend to zero as $n \to \infty$.   So, by the
sandwich result (i) above, conclude that $f(n) \to 0$ as
$n \to \infty$.

55. SEQUENCES WITH OSCILLATORY BEHAVIOUR

<u>Definition</u>. A sequence which is bounded but does not tend to a limit as $n \to \infty$ <u>oscillates boundedly</u>.
[<u>Prototype</u>: $f(n) = (-1)^n$.]

<u>Definition</u>. A sequence which is unbounded but which tends neither to infinity nor to minus infinity <u>oscillates unboundedly</u>. [<u>Prototype</u>: $f(n) = n(-1)^n$.]

<u>EXAMPLE</u>. Let $f(n) = 3$ (n odd), $f(n) = 4$ (n even). <u>Show that</u> $f(n)$ oscillates boundedly as $n \to \infty$.

<u>Solution</u>. Clearly $\{f(n)\}$ is bounded. So either it tends to a limit or it oscillates boundedly. Suppose that it tends to a limit. Then, taking $\varepsilon = 1/4$ in §48(iii) gives the conclusion that

$$\exists X \text{ such that } |f(n_1) - f(n_2)| < \tfrac{1}{2} \quad \forall n_1, n_2 > X.$$

But we can take n_1 even and n_2 odd both greater than X to conclude that $1 < \tfrac{1}{2}$. This is a contradiction. So the assumption is wrong and the sequence oscillates boundedly, as required.

56.
If $f(n) \to A$ as $n \to \infty$, then $|f(n)| \to |A|$ by §51. However, if $|g(n)| \to |B|$ as $n \to \infty$, there is no guarantee that $\{g(n)\}$ even tends to a limit. For example, if $g(n) = (-1)^n$, then $|g(n)| \to 1$ but $\{g(n)\}$ oscillates boundedly.

Notice however the simpler situation that occurs if $|f(n)| \to 0$ as $n \to \infty$. It <u>is</u> true that $f(n) \to 0$ <u>if and only if</u> $|f(n)| \to 0$. This happens because the main clause in the definition, i.e. $|f(n) - 0| < \varepsilon$, is the same each case.

57. MONOTONIC SEQUENCES

Let $\{f(n)\}$ be a sequence. $\{f(n)\}$ is called <u>increasing</u> if $f(n+1) \geq f(n)$ for all $n \varepsilon N$ and <u>strictly increasing</u> if $f(n+1) > f(n)$ for all $n \varepsilon N$. The sequence is <u>increasing for all</u> $n > X$ if $f(n+1) \geq f(n)$ for all $n > X$.

Similarly for <u>decreasing</u> and <u>strictly decreasing</u>. The adjective <u>monotonic</u> means decreasing or increasing.

To show that a sequence $\{f(n)\}$ is increasing, check <u>either</u> that $(f(n+1) - f(n)) \geq 0$, <u>or</u> (provided all the terms are positive) that $f(n+1)/\overline{f(n)} \geq 1$. In any particular problem check whichever is easier.

The following result can be very useful in deciding the behaviour of a monotonic sequence.

RESULT. An increasing sequence is either (i) unbounded above, in which case it tends to infinity, or (ii) bounded above, in which case it tends to a limit. (Likewise for decreasing sequences and bounded below.)

Proof. (i) Suppose that $\{f(n)\}$ is increasing and unbounded above. Then

$$\forall K, \ \exists N \text{ such that } f(N) > K.$$

But then since $\{f(n)\}$ is increasing,

$$f(n) \geq f(N) > K \quad \forall n \geq N.$$

So $f(n) \to \infty$ as $n \to \infty$.

(ii) Suppose that $\{f(n)\}$ is increasing and bounded above. Choose $\varepsilon > 0$. Then $A = \sup_{n \geq 1} f(n)$ exists. Then take $A-\varepsilon$ as σ' in the definition of supremum to deduce the existence of an integer N such that $f(N) > A-\varepsilon$. But then since $\{f(n)\}$ increases

$$A \geq f(n) \geq f(N) > A-\varepsilon \quad \text{for all } n \geq N,$$

i.e. $|f(n)-A| < \varepsilon$ for all $n \geq N$,

i.e. $f(n) \to A$ as $n \to \infty$.

EXAMPLE. A sequence $\{f(n)\}$ is defined on N by $f(n) = \frac{1}{n+1} + \frac{1}{n+2} + \ldots + \frac{1}{2n}$. Prove that $\{f(n)\}$ is increasing. By doing estimation prove that $f(n) \to L$ as $n \to \infty$, where $\frac{1}{2} \leq L \leq 1$.

Solution. (Here it is better to consider $(f(n+1)-f(n))$ rather than $f(n+1)/f(n)$.) For every $n \varepsilon N$, we have

$$f(n+1)-f(n) = \left(\frac{1}{n+2} + \frac{1}{n+3} + \ldots + \frac{1}{2n+1} + \frac{1}{2n+2} \right)$$
$$- \left(\frac{1}{n+1} + \frac{1}{n+2} + \ldots + \frac{1}{2n-1} + \frac{1}{2n} \right).$$
$$= \frac{1}{2n+1} + \frac{1}{2n+2} - \frac{1}{n+1} = \frac{1}{2n+1} - \frac{1}{2n+2} > 0.$$

So $\{f(n)\}$ is increasing.

The expression for $\{f(n)\}$ contains n terms, each at least $\frac{1}{2n}$ and at most $\frac{1}{n+1}$. So $\frac{n}{2n} \leq f(n) \leq \frac{n}{n+1}$ for every $n \varepsilon N$. So $\frac{1}{2} \leq f(n) < 1$ for every $n \varepsilon N$. So $\{f(n)\}$ is increasing and bounded above by 1. Also $f(1) = \frac{1}{2}$. So, by the above result, $f(n) \to L$ as $n \to \infty$, where $\frac{1}{2} \leq L \leq 1$. (In fact $L = \log_e 2 = 0.6931$ (4 places). See §180.)

58. In some sense the following example attempts to answer the meaningless (see §47) question: What is zero times infinity?

EXAMPLE. Give examples to show that if $f(n) \to 0$ and $g(n) \to \infty$, then as $n \to \infty$, $\{f(n)g(n)\}$ may (i) tend to zero, (ii) tend to infinity, (iii) tend to a nonzero limit, (iv) oscillate boundedly, (v) oscillate unboundedly.

Solution. (i) $f(n) = 1/n^2$, $g(n) = n$ and $f(n)g(n) = 1/n$.

(ii) $f(n) = 1/n$, $g(n) = n^2$ and $f(n)g(n) = n$.

(iii) $f(n) = 1/n$, $g(n) = 4n$ and $f(n)g(n) = 4$.

(iv) $f(n) = (-1)^n/n$, $g(n) = n$ and $f(n)g(n) = (-1)^n$.

(v) $f(n) = (-1)^n/n$, $g(n) = n^2$ and $f(n)g(n) = (-1)^n n$.

59. COUNTEREXAMPLES

These answer questions of the type: "Give an example to show that ...". Realise that the definition of a property often starts with the quantifier $\forall$. For example,

$$\forall a,b \in G, \quad ab = ba. \qquad \text{(Group } G \text{ is } \underline{abelian}.)$$

To show that under some conditions the property does not hold, it is enough to give one concrete example for which the result fails. This is enough because the negation of a statement starting with $\forall$ must according to §32 start with $\exists$.
For example,

$$\exists a,b \in G \text{ such that } ab \neq ba. \qquad \text{(} G \text{ is } \underline{non\text{-}abelian}.)$$

Usually a counterexample is demanded to demonstrate that a result fails if one of the conditions for the result is deleted. In attempting to find a counter-example try to exploit the deletion.

Maintain an active vocabulary of examples of different types of behaviour for counterexamples. (See Exs.2.6,2.13,2.14.)

60. LIMITS OF REAL FUNCTIONS AS X TENDS TO INFINITY

For f defined on R (not just on N), $f(x)$ tends to A as x tends to infinity means:

$$\forall \varepsilon > 0, \quad \exists X \text{ such that } |f(x)-A| < \varepsilon \quad \forall x > X.$$

The similarity to the definition for sequences means that much of the above can be carried over directly to real functions.

In particular we can immediately claim analogues of the heredity results (§51), the sandwich theorems (§54) and the result for monotonic sequences (§57). Notice however as mentioned in §49 that a real function which tends to a limit as $x \to \infty$ need <u>not</u> be bounded. The functions $\cos x$ and $x \cos x$ <u>are</u> prototypes of bounded oscillation and unbounded oscillation as $x \to \infty$ respectively.

61. RELATIVE GROWTH OF NTH POWERS AND POWERS OF N

There are many inequalities such as $2^n > n$ and $3^n > 2n^2$ for all $n \epsilon N$, which compare an nth power with a power of n. While particular cases like these can usually be proved by induction, the following result provides a general family of such inequalities.

RESULT. Let $x > 1$ and let $n \epsilon N$. Then for every integer k with $0 \le k \le n$,

$$x^n > \binom{n}{k} (x-1)^k.$$

In particular, $x^n > n(x-1)$.

<u>Proof</u>.
$$x^n = (1+(x-1))^n$$
$$= 1+n(x-1) + \frac{n(n-1)}{2}(x-1)^2 + \ldots$$
$$\ldots + \binom{n}{k}(x-1)^k + \ldots + (x-1)^n.$$

The results follow since each term on the right is positive.

<u>EXAMPLE</u>. Use the above result to show that, for $n \ge 3$,
$$3^n > 2n, \quad 3^n > 2n(n-1) \quad \text{and} \quad 3^n > \frac{4}{3}n(n-1)(n-2).$$

62. BEHAVIOUR OF NTH POWERS AS N TENDS TO INFINITY

<u>RESULT</u>. (i) $a^n \to \infty$ as $n \to \infty$ if $a > 1$;

(ii) $a^n \to 0$ as $n \to \infty$ if $-1 < a < 1$;

and (iii) a^n oscillates unboundedly as $n \to \infty$ if $a < -1$.

<u>Proof</u>. (i) Let $a > 1$. Then, by §61, $a^n > n(a-1)$. As $n \to \infty$, $n(a-1) \to \infty$. So $a^n \to \infty$.

(ii) Let $-1 < a < 1$. Then $1/|a| > 1$. So $(1/|a|)^n \to \infty$ by part (i). So $|a|^n \to 0$, i.e. $a^n \to 0$. So $a^n \to 0$ as $n \to \infty$.

(iii) Let $a < -1$. Then $|a|^n \to \infty$, but a^n alternates in sign. So a^n oscillates unboundedly.

63. BEHAVIOUR OF THE PRODUCT OF AN NTH POWER AND A POWER OF N AS N TENDS TO INFINITY

To discuss the behaviour of $3^n/n$ as $n \to \infty$, we could rearrange the second result of the example in §61 in the form $3^n/n > 2(n-1)$ for all $n \geq 3$. Then since $2(n-1) \to \infty$ as $n \to \infty$, conclude that $3^n/n \to \infty$ as $n \to \infty$. This argument is generalised in the following result, which makes precise the statement that an nth power dominates a power of n as $n \to \infty$.

RESULT. Let $k \varepsilon N$ and let $a > 1$ and $0 < b < 1$. Then, as $n \to \infty$,

$$\text{(i)} \quad a^n/n^k \to \infty, \qquad \text{(ii)} \quad b^n n^k \to 0.$$

Proof. (i) From §61 notice that $a^n > \binom{n}{k+1}(a-1)^{k+1}$ for all $n > k$. So $a^n > \frac{n(n-1)\ldots(n-k)}{(k+1)!}(a-1)^{k+1}$. Rearrange this as

$$\frac{a^n}{n^k} > n(1 - \frac{1}{n})(1 - \frac{2}{n})\ldots(1 - \frac{k}{n})\frac{(a-1)^{k+1}}{(k+1)!}.$$

As $n \to \infty$, the factors $(1 - \frac{r}{n})$ all tend to 1, so that the RHS of the inequality tends to infinity. So $a^n/n^k \to \infty$ as $n \to \infty$.

(ii) Let $c = 1/b$. Then $c > 1$ and by part (i), $c^n/n^k \to \infty$ as $n \to \infty$. So $n^k/c^n \to 0$ as $n \to \infty$, i.e. $b^n n^k \to 0$ as $n \to \infty$.

(For an alternative proof see §65.)

EXAMPLE. Discuss the behaviour as $n \to \infty$ of $\{f(n)\}$, where $f(n) = \frac{n^3+2n+15}{3n+7}\left(\frac{5}{6}\right)^n$ for all $n \varepsilon N$.

Solution. For every $n \varepsilon N$,

$$0 < f(n) \leq \frac{n^3+2n^3+15n^3}{3n}\left(\frac{5}{6}\right)^n = 6n^2\left(\frac{5}{6}\right)^n.$$

By the above result $6n^2\left(\frac{5}{6}\right)^n \to 0$ as $n \to \infty$. So by a sandwich argument $f(n) \to 0$ as $n \to \infty$.

EXAMPLE.   Discuss the behaviour as $n \to \infty$ of $n^4 4^n$, $n^4 4^{-n}, n^{-4} 4^n, n^{-4} 4^{-n}$.

EXAMPLE.   By doing estimation and using a sandwich theorem, discuss the behaviour as $n \to \infty$ of

$$(a) \ \frac{n^4}{4^n - 1}, \qquad (b) \ \frac{n^3 - 4n - 11}{n + 6} \left(\frac{5}{6}\right)^n.$$

EXAMPLE.   Let p be a real polynomial function.   Show that $p(n) 2^{-n} \to 0$ as $n \to \infty$.

EXAMPLE.   Deduce from the above result the behaviour as $n \to \infty$ of $n^k c^n$, where (i) $-1 < c < 0$, (ii) $c < -1$.

EXAMPLE.   Though we demanded that k be a positive integer in the above result, the result is still valid if k is any positive number.   Prove this using the fact that $n^{[k]} \leq n^k < n^{[k]+1}$ $(n \varepsilon N)$, where [ ] denotes the integral part.

OTHER EXAMPLES ON SEQUENCES

64.   RECURSIVE SEQUENCES

Here the (n+1)th term is derived from the nth term in a fixed way.   The following three examples illustrate methods of tackling the convergence of such sequences but there are many variants and each problem must be treated on its merits.   Usually the relation between x_{n+1} and x_n can be used to decide the values of possible limits. To prove convergence it may help to consider $(x_{n+1} - L)$ directly (where L is a possible limit) or to consider $x_{n+1} - x_n$ to see if $\{x_n\}$ is eventually monotonic.   Use of a calculator to find successive terms can provide valuable insight.

(i) <u>EXAMPLE</u>. A sequence $\{x_n\}$ is defined recursively in terms of x_1 by $x_{n+1} = x_n^2 - 8x_n + 20$ $(n \varepsilon N)$. Discuss the behaviour as $n \to \infty$ of $\{x_n\}$ in the cases when (a) $3 < x_1 < 5$, (b) $x_1 > 5$.

<u>Solution</u>. Suppose that $x_n \to L$ as $n \to \infty$. Then $x_{n+1} \to L$ and $x_n^2 - 8x_n + 20 \to L^2 - 8L + 20$ as $n \to \infty$. So $L = L^2 - 8L + 20$. Solution of this quadratic gives the only possible values of limits as $L = 4$ or $L = 5$.

Notice that $(x_{n+1} - 4) = x_n^2 - 8x_n + 16 = (x_n - 4)^2$ for all $n \varepsilon N$. So, in particular, $(x_2 - 4) = (x_1 - 4)^2$ and $(x_3 - 4) = (x_2 - 4)^2 = (x_1 - 4)^{2^2}$ and so on. In fact, by induction, for all $n \varepsilon N$, $(x_{n+1} - 4) = (x_1 - 4)^{2^n}$. In case (a), $(x_1 - 4)^{2^n} \to 0$ as $n \to \infty$ and so $x_{n+1} - 4 \to 0$, i.e. $x_n \to 4$ as $n \to \infty$. In case (b), $(x_1 - 4)^{2^n} \to \infty$ as $n \to \infty$ and so $x_{n+1} - 4 \to \infty$, i.e. $x_n \to \infty$ as $n \to \infty$.

(With $x_1 = 4.80$ a calculator gives successive terms as 4.640, 4.410, 4.168, 4.028, 4.001, 4.000, ... With $x_1 = 5.20$ it gives successive terms as 5.440, 6.074, 8.300, 22.488, 345.822, ...)

(ii) <u>EXAMPLE</u>. A sequence $\{x_n\}$ is defined by letting x_1 be any positive number and $x_{n+1} = (5x_n + 9)^{1/4}$ for every $n \varepsilon N$. Show that the only possible limit of this sequence is a non-negative root of this equation $x^4 - 5x - 9 = 0$. <u>Given</u> that there is only one such root c and that $c \varepsilon \,]2, 3[$, prove that $x_n \to c$ as $n \to \infty$.

<u>Solution</u>. Suppose that $x_n \to L$ as $n \to \infty$. Then $L \geq 0$ because x_n is always positive. Also $x_{n+1} \to L$ and $(5x_n + 9)^{1/4} \to (5L + 9)^{1/4}$ as $n \to \infty$. It follows that $L = (5L + 9)^{1/4}$, i.e. $L^4 - 5L - 9 = 0$. So $L = c$.

Using the fact that $c^4 = 5c + 9$, we can see that

$$x_{n+1}^4 - c^4 = (5x_n + 9) - (5c + 9) = 5(x_n - c).$$

So $|x_{n+1}-c| = \dfrac{5|x_n - c|}{(x_n+c)(x_{n+1}^2+c^2)} \leq \dfrac{5|x_n-c|}{c^3} \leq \dfrac{5}{8}|x_n - c|,$

on using the facts that $x_n \geq 0$ and $c > 2$. It then follows by induction that $|x_{n+1}-c| \leq \left(\dfrac{5}{8}\right)^n |x_1 - c|$, for all $n\epsilon N$. Then, since $(5/8)^n \to 0$ as $n \to \infty$, conclude that $|x_{n+1}-c| \to 0$ as $n \to \infty$, i.e. $x_n \to c$ as $n \to \infty$.

(With $x_1 = 2$ a calculator gives successive terms as 2.0878, 2.0998, 2.1014, 2.1016, 2.1016, ...)

(iii) <u>EXAMPLE</u>. (<u>Newton's method for solving $x^2 = a$</u>) Let $a > 0$ and define the sequence $\{x_n\}$ by taking x_1 as any positive number and then
$$x_{n+1} = (x_n^2 + a)/(2x_n) \qquad (n\epsilon N).$$
Prove that $x_n \geq \sqrt{a}$ for all $n \geq 2$ and deduce that $x_n \to \sqrt{a}$ as $n \to \infty$.

<u>Solution</u>. Suppose that $x_n \to L$ as $n \to \infty$. Then $L \geq 0$ since x_n is clearly always positive. Also $x_{n+1} \to L$ and $(x_n^2 + a)/(2x_n) \to (L^2 + a)/(2L)$.

So $L = (L^2 + a)/(2L)$, i.e. $L^2 = a$. So $L = \sqrt{a}$.

For each $n\epsilon N$, write $x_{n+1} = \frac{1}{2}\left(x_n + (a/x_n)\right)$ and notice that this is greater than or equal to $\sqrt{a}$ by the inequality of the means (§39). So $x_{n+1} \geq \sqrt{a}$ $(n \geq 2)$. (Alternatively you can consider $x_{n+1}-\sqrt{a}$.)

Notice then that $x_{n+1}-x_n = (a - x_n^2)/(2x_n) \leq 0$ for all $n \geq 2$, from what we proved above. So $x_{n+1} \leq x_n$ for all $n \geq 2$, i.e. $\{x_n\}$ is decreasing for $n \geq 2$. Also $\{x_n\}$ is bounded below by 0. So, by the result of §57, conclude that $\{x_n\}$ tends to a limit, which from above must be $\sqrt{a}$. So $x_n \to \sqrt{a}$ as $n \to \infty$.

(With $a = 5$ and $x_1 = 2$ a calculator gives 2.2500, 2.2361, 2.2361, ... as successive terms.)

Notice that in parts (i) and (ii) it may help to remember that $|x_n - L|$ measures the <u>distance</u> between x_n and L on the real line.

65. §64 illustrates how an explicit relation between successive terms of a sequence (i.e. between x_{n+1} and x_n) can be used to find the <u>value</u> of a possible limit. Even if such a relation is not given it may still be possible to find one as in the following example.

<u>EXAMPLE</u>. (<u>§63 revisited</u>) Let $k \varepsilon N$ and let $0 < b < 1$. Let $f(n) = n^k b^n$ $(n \varepsilon N)$. Prove that $f(n) \to 0$ as $n \to \infty$.

<u>Solution</u>. Here $f(n+1)/f(n) = b^{n+1}(n+1)^k/b^n n^k = b\left(1 + \frac{1}{n}\right)^k$.

So, $$f(n+1) = b\left(1 + \frac{1}{n}\right)^k f(n) \qquad (n \varepsilon N). \qquad \dots(*)$$

Now suppose that $f(n) \to L$ as $n \to \infty$. Then letting $n \to \infty$ on both sides in (*) we see that $L = bL$, i.e. $L(1 - b) = 0$. But $b \neq 1$. So $L = 0$.

Notice also that $f(n+1)/f(n) \to b$ as $n \to \infty$. So by §48(i) with a suitably small ε, conclude that there exists X such that

$$0 < \frac{f(n+1)}{f(n)} < 1 \quad \text{for all} \quad n > X.$$

So $\{f(n)\}$ is decreasing for all $n > X$ and it is clearly bounded below by 0. So by §57 $\{f(n)\}$ tends to a limit as $n \to \infty$, and from above the limit must be 0.

<u>EXAMPLE</u>. Let $f(n) = \frac{5.8.11\dots(3n+2)}{1.5.9\dots(4n-3)}$ $(n \varepsilon N)$. Prove that $f(n) \to 0$ as $n \to \infty$.

66. <u>EXAMPLE</u>. Let A be a bounded subset of R that does not contain its supremum σ. Show that there exists a sequence $\{a_n\}$ of points of A such that $a_n \to \sigma$ as $n \to \infty$.

<u>Solution</u>. Take $\sigma' = \sigma - 1$ in the definition of σ and so choose $a_1 \varepsilon A$ with $a_1 > \sigma - 1$. Then take $\sigma' = \sigma - \frac{1}{2}$ and so choose $a_2 \varepsilon A$ with $a_2 > \sigma - \frac{1}{2}$. Continue in this way choosing $a_n > \sigma - \frac{1}{n}$. Then we have

$$\sigma - \frac{1}{n} < a_n < \sigma \qquad (n \varepsilon N).$$

Use a sandwich argument on this relation to show that $a_n \to \sigma$ as $n \to \infty$.

EXAMPLES 2

(All numbers in these examples are real.)

1. Sequences $\{f(n)\}$ and $\{g(n)\}$ are defined by

$$f(n) = \frac{6n^2-n}{2n^2-5n+4}, \qquad g(n) = \frac{n^3-5n^2}{n^2+n+1}.$$

Prove from the definitions that $f(n) \to 3$ and $g(n) \to \infty$ as $n \to \infty$.

2. Evaluate (using heredity results):

(a) $\lim\limits_{n\to\infty} \dfrac{3n^2-5n+7}{n^2-2n+5}$,   (b) $\lim\limits_{n\to\infty} \dfrac{3n^3+(-1)^n n}{4n^3-5}$,

(c) $\lim\limits_{n\to\infty} \dfrac{2n+1}{n^3+n+1}$.

3. For real numbers $a_1, a_2, \ldots, a_n$ notice that
$$n \min(a_1, a_2, \ldots, a_n) \le a_1+a_2+\ldots+a_n \le n \max(a_1, a_2, \ldots, a_n).$$
Use this result to show that

$$\frac{1}{n+1} \le \frac{1}{n^2+1} + \frac{1}{n^2+2} +\ldots+ \frac{1}{n^2+n} \le \frac{1}{n}.$$

Deduce that

$$\lim_{n\to\infty} \frac{1}{n^2+1} + \frac{1}{n^2+2} +\ldots+ \frac{1}{n^2+n} = 0.$$

4. Let $f(n) = \dfrac{1}{\sqrt{(n^2+1)}} + \dfrac{1}{\sqrt{(n^2+2)}} +\ldots+ \dfrac{1}{\sqrt{(n^2+n)}}$

for all $n\epsilon N$.   Prove that for all $n\epsilon N$

$$\frac{\sqrt{n}}{\sqrt{(n+1)}} \le f(n) \le \frac{n}{\sqrt{(n^2+1)}}.$$

Deduce that $f(n) \to 1$ as $n \to \infty$.

5. Use heredity results to prove that if $f(n) \to A$ and $g(n) \to B$ as $n \to \infty$, then $\max(f(n),g(n)) \to \max(A,B)$ and $\min(f(n),g(n)) \to \min(A,B)$ as $n \to \infty$.

6. Illustrate each of the following possibilities as $n \to \infty$ by giving an example of a suitable sequence $\{a_n\}$:

(a) $\{|a_n|\}$ tends to a limit, but $\{a_n\}$ does not;

(b) $\{a_n^2\}$ tends to a limit, but $\{a_n\}$ does not;

(c) the sequence $\{a_2, a_4, a_6, \ldots\}$ tends to a limit, but $\{a_n\}$ does not;

(d) $\{a_n a_{n+1}\}$ tends to a limit, while $\{a_n\}$ oscillates boundedly;

(e) $\{a_n\}$ tends to zero, but is not eventually monotonic;

(f) $\{a_n\}$ tends to zero, but $\{\frac{1}{\sqrt{n}}(a_{n+1}+a_{n+2}+\ldots+a_{2n})\}$ does not tend to zero.

7. For the positive sequence $\{f(n)\}$ it is given that there exist numbers X and α such that

$$0 < \frac{f(n+1)}{f(n)} < \alpha < 1 \quad \text{for all} \quad n > X.$$

Prove that $\{f(n)\}$ is eventually decreasing and that $f(n) \to 0$ as $n \to \infty$.

Give an example to show that if the condition above is relaxed and we demand only that

$$0 < \frac{f(n+1)}{f(n)} < 1 \quad \text{for all} \quad n > X,$$

then we <u>cannot</u> conclude that $f(n) \to 0$ as $n \to \infty$.

8. The sequence $\{f(n)\}$ is defined by $f(n) = \frac{2.4.6\ldots(2n)}{1.3.5\ldots(2n-1)} \cdot \left(\frac{3}{4}\right)^n$. Prove that $f(n+1)/f(n) \to \frac{3}{4}$ as $n \to \infty$ and deduce that $f(n) \to 0$ as $n \to \infty$.

9. The sequence $\{a_n\}$ is defined by

$$a_n = \frac{1}{2n+1} + \frac{1}{2n+3} + \ldots + \frac{1}{4n-1}.$$

Prove that $\{a_n\}$ is an increasing sequence. By doing some rough inequality work, prove that $\frac{1}{4} \leq a_n \leq \frac{1}{2}$ for all $n \epsilon N$. Deduce that $a_n \to L$ as $n \to \infty$, where $\frac{1}{4} \leq L \leq \frac{1}{2}$.

10. Suppose that $f(n) \to \frac{1}{2}$ as $n \to \infty$. Show that there exists a number X such that $\frac{1}{4} < f(n) < \frac{3}{4}$ for all $n > X$. Deduce that $(f(n))^n \to 0$ as $n \to \infty$.

Make statements generalising this result, i.e. for a sequence $\{f(n)\}$ such that $f(n) \to A$ as $n \to \infty$, make statements about the behaviour of $\{(f(n))^n\}$ as $n \to \infty$.

(Deal with the particular cases $|A| < 1$, $A > 1$, $A < -1$. For some insight into the case $A = 1$, see §214.)

11. Discuss the behaviour of $n^n a^n$ as $n \to \infty$, where $a > 0$.

12. Discuss the behaviour as $n \to \infty$ of

$$\text{(a)} \quad (n^3+n) \left(\frac{1}{2}\right)^n, \qquad \text{(b)} \quad 2^n (n^3+n)^{-1},$$

$$\text{(c)} \quad \frac{2n^3+1}{n+1} \left(\frac{3}{4}\right)^n, \qquad \text{(d)} \quad \frac{2^{2n}+n}{n^3 3^n+1}, \qquad \text{(e)} \quad \sqrt{(n+1)} - \sqrt{n}.$$

13. Illustrate each of the following possibilities as $n \to \infty$, by giving examples of suitable sequences $\{a_n\}$ and $\{b_n\}$:

(a) $a_n \to \infty$, $b_n \to -\infty$ and $a_n + b_n \to \infty$;

(b) $\{a_n b_n\}$ tends to a limit, while $\{a_n\}$ and $\{b_n\}$ both oscillate;

(c) $\{a_n + b_n\}$ tends to a limit, while neither $\{a_n\}$ nor $\{b_n\}$ does;

(d) $a_n \to 0$, $b_n \to 0$ and $a_n/b_n \to \infty$;

(e) $a_n/b_n \to 1$ as $n \to \infty$, but $(a_n - b_n) \to \infty$.

14. Illustrate each of the following possibilities as $n \to \infty$, by giving an example of a suitable sequence $\{a_n\}$:

(a) $\{a_n\}$ tends to infinity, but is not increasing from some point onwards;

(b) $\{a_n\}$ is positive, tends to zero, but is not decreasing from some point onwards;

(c) $a_{n+1}/a_n \to 1$, but $(a_{n+1} - a_n) \to \infty$;

(d) $a_{n+1}/a_n \to -1$, but $\{a_n\}$ does tend to a limit;

(e) $\{a_n\}$ tends to a limit, but a_{n+1}/a_n oscillates boundedly;

(f) $a_n \to 0$, but $\frac{1}{\sqrt{n}} (a_{n+1} + a_{n+2} + \ldots + a_{2n}) \to \infty$;

(g) $\{a_n\}$ is a positive sequence, but the sequences $\{a_n\}, \{na_n\}, \{n^2 a_n\}, \{n^3 a_n\}, \ldots$ all tend to zero.

15. Prove that if $\{f(n)\}$ oscillates boundedly and $\{g(n)\}$ tends to zero as $n \to \infty$, then $\{f(n)g(n)\}$ tends to zero as $n \to \infty$.

16. If $\{f(n)\}$ oscillates boundedly and $\{g(n)\}$ tends to a non-zero limit as $n \to \infty$, prove by a contradiction argument that $\{f(n)g(n)\}$ oscillates boundedly as $n \to \infty$.

17. Prove that a sequence $\{a_n\}$ with the property that $(a_{n+1}-a_n) \geq 1$ for all $n\epsilon N$ cannot tend to a limit as $n \to \infty$.

18. $\{a_n\}$ is a _positive_ sequence which tends to zero as $n \to \infty$. Prove that $\sup\limits_{n \geq 1} a_n$ is attained, (i.e. there exists $k\epsilon N$ such that $a_k = \sup\limits_{n \geq 1} a_n$).

19. Let $\{a_n\}$ be a sequence of non-negative numbers. Prove that if $a_n \to L$ as $n \to \infty$, then $\sqrt{a_n} \to \sqrt{L}$. (This result is not covered by the heredity results of §51.)

20. If $a_n^3 \to L^3$ as $n \to \infty$, prove that $a_n \to L$.

21. A sequence $\{a_n\}$ is defined recursively in terms of a_1 by
$$a_{n+1} = a_n^2 - 2a_n + 2 \quad (n\epsilon N).$$
Show that if $\{a_n\}$ does tend to a limit, then the limit is 1 or 2. By considering $(a_{n+1}-1)$ or otherwise, prove that

(a) if $0 < a_1 < 2$, then $a_n \to 1$ as $n \to \infty$,

(b) if $a_1 < 0$ or $a_1 > 2$, then $a_n \to \infty$ as $n \to \infty$.

Discuss the cases $a_1 = 0$ and $a_1 = 2$.

22. A sequence $\{a_n\}$ is defined recursively in terms of a_1 by
$$a_{n+1} = a_n^2 - 6a_n + 12 \quad (n\epsilon N).$$
Show that the only possible limits of $\{a_n\}$ are 3 and 4. By evaluating $(a_{n+1}-3)$ and $(a_{n+1}-4)$ in terms of a_n, show that if $2 < a_1 < 4$, then $a_n \to 3$ as $n \to \infty$. Show also that if $a_1 > 4$, then $a_n \to \infty$ as $n \to \infty$. Show also that if $a_1 < 2$, then $a_2 > 4$ and that then $a_n \to \infty$ as $n \to \infty$. What happens if $a_1 = 2$ or $a_1 = 4$?

23. A sequence $\{a_n\}$ is defined by taking a_1 as an arbitrary positive number and then defining
$$a_{n+1} = \sqrt{(a_n + 12)} \quad (n\epsilon N).$$

Demonstrate on a calculator that $a_n \to 4$ as $n \to \infty$. Prove that $a_{n+1}^2 - 16 = a_n - 4$. Deduce that $|a_{n+1} - 4| \leq \frac{1}{4}|a_n - 4|$ and hence prove that $a_n \to 4$ as $n \to \infty$.

24. A sequence $\{a_n\}$ is defined by taking a_1 as an arbitrary positive number and then defining

$$a_{n+1} = (a_n + 5)^{1/4} \qquad (n \varepsilon N).$$

Prove that the only possible limit of this sequence is the positive root α of the equation $x^4 - x - 5 = 0$. (It is given that there is only one such root.)
   Evaluate $a_{n+1}^4 - \alpha^4$ and hence prove that $a_n \to \alpha$ as $n \to \infty$.
Use a calculator to show that $\alpha = 1.60301$ (5 places).

25. Prove that a bounded real sequence which attains neither its supremum nor its infimum oscillates boundedly as $n \to \infty$.

26. It is given that $a_n \to L$ as $n \to \infty$, but that for no integer n is $a_n = L$. Prove that there are infinitely many integers N such that $|a_{N+1} - L| < |a_N - L|$.

27. Let $0 < a < 1$ and let f and g be defined on N by

$$f(n) = n!a^n \quad \text{and} \quad g(n) = n!a^{n^2}.$$

Show that $f(n+1)/f(n) \to \infty$ and $g(n+1)/g(n) \to 0$ as $n \to \infty$. Deduce the behaviour of $\{f(n)\}$ and $\{g(n)\}$ as $n \to \infty$.

28. A sequence $\{a_n\}$ is defined recursively in terms of a_1 by $a_{n+1} = a_n^2 - 8a_n + 14$ $(n \varepsilon N)$. Prove that this sequence converges if and only if $a_n = 2$ or 7 for some value of n.

29. If $a_n \to L$ as $n \to \infty$, prove that $\frac{1}{n}(a_1 + a_2 + \ldots + a_n) \to L$ as $n \to \infty$.

30. Prove that if $f(n) \to \infty$ and $\{g(n)\}$ oscillates boundedly as $n \to \infty$, then $\{f(n)g(n)\}$ cannot be bounded.

31. Discuss the behaviour as $n \to \infty$ of $2^{2^n}/5^n$.

32. Prove that a real sequence $\{f(n)\}$ cannot tend to two different limits as $n \to \infty$. (In a general topological space this need not be true.)

III. LIMITS OF REAL FUNCTIONS AS $x \to c$, CONTINUITY

67. INTERVALS

]1,2[,[1,2],]1,2],[1,2[,[1,∞[and]-∞,∞[are all intervals in R. The first is <u>open</u> and the second is <u>closed</u>; the first four are <u>bounded</u> and the last two <u>unbounded</u>.

68. NEIGHBOURHOOD

Let $c \epsilon R$ and $U \subseteq R$. U is a <u>neighbourhood of c</u> if there exists an <u>open</u> interval I such that $c \epsilon I$ and $I \subseteq U$. For example,]1,3[,[1,∞[and [1,5] are all neighbourhoods of 2 but [2,3[is not.

Let $N(c,\delta)$ denote]c-δ,c+δ[. Let N'(c,δ) denote]c-δ,c[∪]c,c+δ[. Notice that $N(c,\delta)$ is a neighbourhood of c, but that N'(c,δ) is not since it fails to contain c.

69. LIMIT AS $x \to c$

<u>Definition</u>. For a real function f defined on a neigh-bourhood of a point c, the statement that <u>f(x) tends to A as x tends to c</u> means:

$\forall \epsilon > 0$, $\exists \delta > 0$ such that $|f(x)-A| < \epsilon$ $\forall x \epsilon N'(c,\delta)$.

[Write: $f(x) \to A$ as $x \to c$ or $\lim\limits_{x \to c} f(x) = A$.]

Notice: (i) As in §41, ε is chosen first and it then dictates the value of δ. In general, the smaller ε is chosen, the smaller δ must be taken, i.e. to get the values f(x) closer to A, you must take the values of x closer to c.

(ii) The behaviour of f <u>at the point</u> c does not affect the existence or non-existence of the limit. For example $\lim\limits_{x \to 0}$ (sin x)/x = 1, even though (sin x)/x is undefined at x = 0. Look also at §70(ii).

(iii) f(x) <u>tends to A as x tends to c on the right</u> means:

$\forall \epsilon > 0$, $\exists \delta > 0$ such that $|f(x)-A| < \epsilon$ $\forall x \epsilon]c,c+\delta[$.

[Write: $\lim\limits_{x \to c+} f(x) = A$ or f(c+) = A.]

Similarly for limits <u>on the left</u>, denoted by $\lim\limits_{x \to c-} f(x)$ or f(c-).

70. <u>EXAMPLE</u>. The function f is defined on R by

$$f(x) = \begin{cases} 4x+6 & (x > 0), \\ 23 & (x = 0), \\ -2x+6 & (x < 0). \end{cases}$$

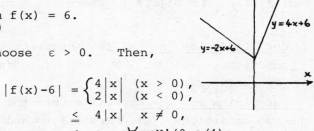

Show that $\lim_{x \to 0} f(x) = 6.$

<u>Solution</u>. Choose $\varepsilon > 0.$ Then,

$$|f(x)-6| = \begin{cases} 4|x| & (x > 0), \\ 2|x| & (x < 0), \end{cases}$$

$$\leq\ 4|x| \quad x \neq 0,$$

$$<\ \varepsilon \quad \forall x \varepsilon N'(0,\varepsilon/4).$$

So $\forall \varepsilon > 0,\ |f(x)-6|\ <\ \varepsilon \quad \forall x \varepsilon N'(0,\varepsilon/4).$

Notice: (i) Here $\delta = \varepsilon/4$ and δ depends on ε in the same way as X depends on ε in §41.   Notice that as ε becomes smaller, so does δ.

(ii) The existence and value of the limit are unaffected by the value of f(0).

71. <u>EXAMPLE</u>. The function g is defined on R by

$$g(x) = \begin{cases} 4x+6 & (x \geq 0), \\ -2x+1 & (x < 0). \end{cases}$$

Find g(0+) and g(0-).

<u>Solution</u>. Choose $\varepsilon > 0.$

<u>For x > 0</u>, $|g(x)-6| = 4|x| < \varepsilon \quad \forall x \varepsilon]0, \varepsilon/4[.$

<u>For x < 0</u>, $|g(x)-1| = 2|x| < \varepsilon \quad \forall x \varepsilon]-\varepsilon/2, 0[.$

So g(0+) = 6 and g(0-) = 1.

Notice also that $\lim_{x \to 0} g(x)$ does <u>not</u> exist, even although both one-sided limits exist at 0.

72. The examples of §70 and §71 illustrate the following result.

<u>RESULT</u>. $\lim_{x \to c} f(x) = A \Longleftrightarrow f(c+)$ and f(c-) both exist and are equal to A.

<u>Proof</u>. $\Longrightarrow$ is clear because $|f(x)-A| < \varepsilon\ \forall x \varepsilon N'(c,\delta)$ means $|f(x)-A| < \varepsilon \quad \forall x \varepsilon]c,c+\delta[$ <u>and</u> $\forall x \varepsilon]c-\delta,c[.$

To prove $\Leftarrow$, suppose that $f(c+) = f(c-) = A$.
Then $|f(x)-A| < \varepsilon$ $\forall x \varepsilon]c,c+\delta_1[$ and $|f(x)-A| < \varepsilon$
$\forall x \varepsilon]c-\delta_2,c[$. It follows that

$|f(x)-A| < \varepsilon$ $\forall x \varepsilon N'(c,\delta_3)$, where $\delta_3 = \min(\delta_1,\delta_2)$.

73. For limits as $n \to \infty$, the intersection of the two
conditions (α) $\forall n > X_1$ and (β) $\forall n > X_2$ is
(γ) $\forall n > \max(X_1,X_2)$.

For limits as $x \to c$, notice that the intersection
of the two conditions (α) $\forall x \varepsilon N(c,\delta_1)$ and (β) $\forall x \varepsilon N(c,\delta_2)$
is (γ) $\forall x \varepsilon N(c,\delta_3)$, where $\delta_3 = \min(\delta_1,\delta_2)$.

74. HEREDITY AND SANDWICH RESULTS, LIMITS OF MONOTONIC
FUNCTIONS

For limits as $x \to c$ or $x \to c+$, heredity results
(on sums and products etc.) and sandwich results are
easily proved along similar lines to §51 and §54. Also,
analogues of the results on monotonic sequences (§57) can
be proved for one-sided limits, e.g. If f is increasing
and bounded above on $[a,c[$ then $\lim_{x \to c-} f(x)$ exists.

75. THE BEHAVIOUR OF POLYNOMIALS AND RATIONAL FUNCTIONS
AS X TENDS TO C

Notice that trivially $x \to c$ as $x \to c$, because

$$\forall \varepsilon > 0, \quad |x-c| < \varepsilon \quad \forall x \varepsilon]c-\varepsilon,c+\varepsilon[.$$

Now use the heredity results mentioned in §74 with the
function $\{x\}$ as a building block. First use products
to conclude that $x^2 \to c^2$ and indeed $x^n \to c^n$ ($n\varepsilon N$).
Now use addition and scalar multiplication to conclude
that $p(x) \to p(c)$ as $x \to c$ for every real polynomial p.

Further, take the quotient of two real polynomials
p and q to deduce that $p(x)/q(x) \to p(c)/q(c)$ as $x \to c$,
provided that $q(c) \neq 0$. So, for every real rational
function r, $r(x) \to r(c)$ as $x \to c$, provided that $r(c)$
is defined.

EXAMPLE.   Give a proof (independent of the above
argument that $x^3 \rightarrow 8$ as $x \rightarrow 2$.   [Hint. Estimate
$|x^3-8|$ as in Ex.1.3(ii).]

76.   Definitions of $f(x) \rightarrow \pm\infty$ as $x \rightarrow c$ are analogous
      to those in §45.   The following definition
illustrates this.

Definition. $f(x) \rightarrow \infty$ as $x \rightarrow c+$ means:

$\forall K, \exists \delta > 0$ such that $f(x) > K$ $\forall x \epsilon]c,c+\delta[.$

(Recall that without loss of generality we can take $K > 0$
as in §45.)

EXAMPLE.   Prove from the definition that f defined on
$R-\{-5,7\}$ by
$$f(x) = \frac{x}{(x+5)(x-7)}$$
tends to infinity as $x \rightarrow 7+$. (Compare §38.)

77.   OSCILLATION AS X TENDS TO C
        As $x \rightarrow c$, some functions neither tend to a limit
as in §69 nor tend to $\pm\infty$ as in §76.   They oscillate
boundedly or unboundedly: respective prototypes are $\cos\frac{1}{x}$
and $\frac{1}{x}\cos\frac{1}{x}$ as $x \rightarrow 0$.

EXAMPLE.   Show that f defined by $f(x) = \cos\frac{1}{x}$ $(x \neq 0)$
oscillates boundedly as $x \rightarrow 0$.

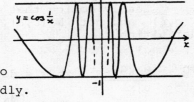

Solution.   The graph of this
function shows rapid oscillation
near $x = 0$. f is bounded on
$R-\{0\}$ and so either f tends to
a limit A or oscillates boundedly.
Suppose then that $f(x) \rightarrow A$ as $x \rightarrow 0$.   Then by taking
$\epsilon = 1/2$ in the definition deduce that
    $\exists \delta > 0$ such that $|f(x)-A| < 1/2$ $\forall x \epsilon N'(0,\delta)$.
So for all pairs of points $x_1, x_2 \epsilon N'(0,\delta)$,
$$|f(x_1)-f(x_2)| \leq |f(x_1)-A| + |A-f(x_2)|$$
$$< 1/2 + 1/2$$
$$= 1. \qquad \ldots (*)$$

Choose a positive integer N such that $1/(2N\pi) < \delta$.
Let $x_1 = 1/(2N\pi)$ and $x_2 = 1/((2N+1)\pi)$. Then
$x_1, x_2 \epsilon N'(0,\delta)$, but $|f(x_1)-f(x_2)| = |1-(-1)| = 2$. This
contradicts (*). So f(x) does not tend to a limit as
$x \to 0$ and so f oscillates boundedly as $x \to 0$.

Notice the similarity between §48(iii) and the
method used to get (*).

CONTINUITY

78.   If $\lim\limits_{x \to c} f(x)$ exists, it need not in general be
equal to f(c). §70 illustrates this. However,
if the two values are in fact equal, we then say that f
is <u>continuous at the point c</u>, i.e. if $\lim\limits_{x \to c} f(x) = f(c)$.

<u>Definition</u>. The real function f is <u>continuous at   c</u>
means that $\lim\limits_{x \to c} f(x) = f(c)$; i.e.

$\forall \epsilon > 0, \exists \delta > 0$ such that $|f(x)-f(c)| < \epsilon \ \forall x \epsilon N(c,\delta)$.

In general terms f is continuous at c if the
graph of f does not jump at the point (c,f(c)).   §79
and §80 illustrate the definition.
Many well-known functions (e.g. polynomials) are
continuous at all points of R or at all points of some
other interval. Accordingly we say that f is
<u>continuous on the interval I</u> if f is continuous at
every point of I.
Lastly notice that (for trivial reasons) it is
$N(c,\delta)$ and not $N'(c,\delta)$ that appears in the above
definition.

79.   <u>EXAMPLE</u>. h is defined on R by

$$h(x) = \begin{cases} 4x+6 & (x \geq 0), \\ -2x+6 & (x < 0). \end{cases}$$

Show that h is continuous at 0.

<u>Solution</u>. Choose $\epsilon > 0$. Then

$$|h(x)-h(0)| = |h(x)-6| = \begin{cases} 4|x| & (x \geq 0), \\ 2|x| & (x < 0), \end{cases}$$
$$\leq \ 4|x| \quad \forall x \epsilon R,$$
$$< \ \epsilon \quad \forall x \epsilon N(0,\epsilon/4).$$

So $\forall \varepsilon > 0$, $|h(x)-h(0)| < \varepsilon$ $\forall x \varepsilon N(0,\varepsilon/4)$. So h is
continuous at 0.

80. DISCONTINUITIES

Neither f in §70 nor g in §71 is continuous
at 0.

For f, $\lim\limits_{x \to 0} f(x) = 6 \neq f(0)$.

For g, $\lim\limits_{x \to 0} g(x)$ does not even exist, let alone
being equal to g(0). The graphs of f and g each
make some sort of jump at x = 0, corresponding to the
discontinuity.

81. f is continuous on the right at c means that
$\lim\limits_{x \to c+} f(x) = f(c)$, i.e.

$\forall \varepsilon > 0$, $\exists \delta > 0$ such that $|f(x)-f(c)| < \varepsilon$ $\forall x \varepsilon [c,c+\delta[$.
Similarly on the left.

From §72, continuity at c is equivalent to
continuity on both the right and left at c.

82. INHERITANCE OF CONTINUITY

The heredity results for limits as $x \to c$ mentioned
in §74 give heredity results for functions continuous at c.

RESULT. Let f and g be real functions that are
continuous at c and let kεR. Then kf,$|f|$,f+g,fg,
max(f,g) and min(f,g) are all continuous at c. So
also is f/g provided that g(c) $\neq$ 0.

Proof. For example, using the result for limits in §74,
we see that

$$\lim_{x \to c} (fg)(x) = \lim_{x \to c} f(x)g(x) = \lim_{x \to c} f(x).\lim_{x \to c} g(x)$$
$$= f(c)g(c) \quad \text{by the continuity of f and g,}$$
$$= (fg)(c).$$

83. INHERITANCE OF CONTINUITY UNDER COMPOSITION

The composition $f_{\bullet}g$ of two functions f, g is
defined by $(f_{\bullet}g)(x) = f(g(x))$. [N.B. The product fg
is defined by (fg)(x) = f(x)g(x).] According to
the following result, continuity is also inherited

under composition.   Later this lets us conclude that, for example, $\sin(x^3+1)$ and $\log \cosh x$ are continuous on R.

RESULT.   Let g be continuous at c and let f be continuous at b, where b = g(c).   Then $f_{\mathbf{o}}g$ is continuous at c.

Proof.

$\forall \epsilon > 0, \quad \exists \delta_1 > 0$ such that $|f(y)-f(b)| < \epsilon$
$$\forall y \epsilon]b-\delta_1, b+\delta_1[. \quad \ldots(1)$$
Now take ϵ in the definition of continuity of g to be δ_1.   So, as in §48(ii), find δ_2 such that
$$g(x)\epsilon]b-\delta_1, b+\delta_1[\quad \forall x\epsilon]c-\delta_2, c+\delta_2[.$$
So g(x) can play y in (1).   So
$$\forall \epsilon > 0, \quad \exists \delta_2 > 0 \quad \text{such that} \quad |f(g(x))-f(g(c))| < \epsilon$$
$$\forall x\epsilon]c-\delta_2, c+\delta_2[.$$
So $f_{\mathbf{o}}g$ is continuous at c.

Variants of this result are similar.   For example:

If $g(x) \to b$ as $x \to \infty$, and f is continuous at b, then $(f_{\mathbf{o}}g)(x) \to f(b)$ as $x \to \infty$.

84.   CONTINUITY OF POLYNOMIALS AND RATIONAL FUNCTIONS

Every real polynomial is continuous on R.   Every real rational function is continuous wherever it is defined (i.e. except at the zeros of the denominator). These are the results of §75.

Later we see that many other functions like $\sin x$, $\sinh x$ and e^x are also continuous on R.

EXAMPLE.   Prove from the definition of continuity that x^2-5x+8 is continuous at x = 3.

85.   EXAMPLE.   A function f is defined on R by

$$f(x) = \begin{cases} 4x^2 & (x \text{ rational}), \\ -2x^2 & (x \text{ irrational}). \end{cases}$$

Show that (i) f is continuous at x = 0, (ii) f is not continuous at x = 1.

53

Solution.   (i) Choose $\epsilon > 0$.   Then

$$|f(x)-f(0)| = \begin{cases} 4x^2 & (x \text{ rational}), \\ 2x^2 & (x \text{ irrational}). \end{cases}$$

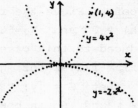

$$\leq\ 4x^2\ \forall x\epsilon R,$$
$$<\ \epsilon\ \ \forall x\epsilon N(0,\tfrac{1}{2}\sqrt{\epsilon}).$$

So f is continuous at 0.

   (ii) Suppose that f is continuous at $x = 1$. ...(α)
Take $\epsilon = 1/2$ in the definition to get
   $\exists\ \delta > 0$ such that $|f(x)-f(1)| < 1/2\ \forall x\epsilon N(1,\delta)$,
            i.e. $|f(x)-4| < 1/2\ \forall x\epsilon N(1,\delta)$.
So $f(x) > 7/2$ for all $x\epsilon N(1,\delta)$.   ...(β)

   Now choose an irrational point $a\epsilon]1,1+\delta [$.   Then
$f(a) = -2a^2 < -2$.   But $f(a) > 7/2$ by (β).
Contradiction.   So the supposition (α) must be false.
So f is <u>not</u> continuous at 1.

   Notice that here f is continuous at 0, though its
graph is not a continuous curve on a neighbourhood of 0.
   Notice also that in (i), the value $\tfrac{1}{2}\sqrt{\epsilon}$ for δ is
found by solving the inequality $4x^2 < \epsilon$.

86. THE USE OF A CONTRADICTION ARGUMENT

   This often has success in proving that a function
is <u>not</u> continuous at a point (as in §85), or does <u>not</u>
tend to a limit (as in §55 and §77) or does <u>not</u> tend to
infinity (as in the example below).   Begin the proof by
supposing not and making a particular choice of ϵ or K.

<u>EXAMPLE</u>.   The real function f is defined on R and is
periodic with period 1 (i.e. $f(x+1) = f(x)$ for all $x\epsilon R$).
Prove that f cannot tend to infinity as $x \to \infty$.
[<u>Hint</u>: Suppose not and let $K = f(0)+1$.]

87. THE INTERMEDIATE VALUE THEOREM

   This result shows that the graph of a function that
is continuous <u>on an interval</u> must effectively be a
"continuous curve" on that interval.

<u>RESULT</u>.   Let f be continuous on [a,b] and let K be
a real number between $f(a)$ and $f(b)$.   Then there exists
a point $c\epsilon [a,b]$ such that $f(c) = K$.

<u>Proof</u>.   Prove the result for the case where $f(a) < 0$,
$f(b) > 0$ and $K = 0$.   (The general case is easily
reduced to this case by looking at g where
$g(x) = f(x)-K$ or $g(x) = K-f(x)$.)

Let $S = \{y: y\epsilon[a,b], f(y) < 0\}$.   Now $a\epsilon S$ so that
$S \neq \phi$ and S is bounded above by b.   So $c = \sup S$
exists.

Now either c is or is not a member of S.   If it
is, then $f(c) < 0$.   If it is not, use §66 to conclude
the existence of a sequence of points $c_n\epsilon S$ such that
$c_n \rightarrow c$ as $n \rightarrow \infty$.   But then for every n it follows
that $f(c_n) < 0$ and by the continuity of f it follows
that $f(c) = \lim_{n\to\infty} f(c_n) \leq 0$.   So, in both cases, we know
that $f(c) \leq 0$.   Also it follows that $c < b$.

Also however, $f(c) = \lim_{x\to c+} f(x) \geq 0$, because $f(x) \geq 0$
for all $x > c$ and f is continuous.

So conclude that $f(c) = 0$ as required.

Notice the following points about the above result:

(i) It is a particular case of the result that in a
general metric space connectedness is preserved under the
action of a continuous function.

(ii) It can help in locating the roots of a polynomial
equation.   See §132,§135(ii),Exs.3.10,3.11,5.13,etc.

88.   <u>EXAMPLE</u>. f is a continuous mapping of $[a,b]$
into $[a,b]$.   Prove that there exists a point
$c\epsilon[a,b]$ such that $f(c) = c$.

<u>Solution</u>.   Let $g(x) = f(x)-x$ for all $x\epsilon[a,b]$.   If
$g(a) = 0$ or $g(b) = 0$, the result holds.   Otherwise it
follows from the given conditions that $g(a) > 0$, $g(b) < 0$
and g is continuous on $[a,b]$.   So, by the intermediate
value theorem, there exists $c\epsilon[a,b]$ such that $g(c) = 0$,
i.e. $f(c) = c$.

This is a particular case of the Brouwer fixed point
theorem, c being the <u>fixed point</u>, (i.e. c is not shifted
by the mapping).   A two-dimensional analogue of the same
result applies if a map of London is laid on the ground
somewhere in London.   Then a point on the map coincides
with the point on the ground that it represents.

89. CONTINUITY OF INVERSE FUNCTIONS

Consider a strictly increasing continuous function
$f: [a,b] \to R$.   For all $x \varepsilon [a,b]$, $f(a) \le f(x) \le f(b)$
since f is strictly increasing.   Also, for all
$y \varepsilon [f(a),f(b)]$, there exists $x \varepsilon [a,b]$ such that $f(x) = y$,
by the intermediate value theorem since f is continuous.
So the range of f is $[f(a),f(b)]$.

So conclude that $f: [a,b] \to [f(a),f(b)]$ is a
bijective continuous mapping.   The existence of an inverse
mapping $f^{-1}: [f(a),f(b)] \to [a,b]$ is then guaranteed.
That f^{-1} is strictly increasing is fairly clear.   That
f^{-1} is continuous is proved in the following result.

RESULT.   In the above situation the mapping
$f^{-1}: [f(a),f(b)] \to [a,b]$ is continuous.

Proof.   We prove continuity at s where $s \varepsilon]f(a),f(b)[$.
(The endpoints can be done similarly.)

Let $c = f^{-1}(s)$.   Choose $\varepsilon > 0$ sufficiently small
that $a < c-\varepsilon < c+\varepsilon < b$.   Then let $f(c-\varepsilon) = s-\delta_1$ and
$f(c+\varepsilon) = s+\delta_2$, where δ_1 and δ_2 are positive.   Let
$\delta = \min(\delta_1,\delta_2)$.   Then for all $y \varepsilon]s-\delta,s+\delta[$,

$$f^{-1}(s-\delta_1) < f^{-1}(y) < f^{-1}(s+\delta_2),$$

i.e. $f^{-1}(s) - \varepsilon < f^{-1}(y) < f^{-1}(s) + \varepsilon$,

i.e. $|f^{-1}(y)-f^{-1}(s)| < \varepsilon$ for all $y \varepsilon]s-\delta,s+\delta[$.

So f^{-1} is continuous at s.

Notice: (i) Similar results apply for strictly
decreasing continuous functions and for other types of
interval.

(ii) An analogous result applies to a bijective
continuous mapping between two general metric spaces.

90. FRACTIONAL INDICES

Look at the function $f: [0,\infty[\to [0,\infty[$ defined by
$f(x) = x^n$ where $n \varepsilon N$.   The range of f is $[0,\infty[$ and

f is strictly increasing and continuous. So by the results of §89 (for an infinite interval), there exists a strictly increasing inverse function g: $[0,\infty[\to [0,\infty[$ defined by

$$g(y) = x \iff y = x^n.$$

In this situation we define $y^{1/n}$ by $y^{1/n} = g(y)$ for $y \geq 0$ and $n \varepsilon N$. Then define $y^{m/n}$ by $y^{m/n} = (y^{1/n})^m$ for $m \varepsilon Z$.

Rational indices so defined obey the normal index laws. This can be checked here but will emerge later in §210.

Notice that if n is <u>odd</u>, the definition of $y^{1/n}$ can be extended to negative y. For example f, defined by $f(x) = x^3$, is strictly increasing not just on $[0,\infty[$ but in fact on R. This allows us to get unique real nth roots of all real numbers provided that n is an <u>odd</u> positive integer. On the other hand, if n is <u>even</u>, x^n is not strictly increasing on R and attempts to define nth roots of negative numbers run into trouble. For example, notice the following contradiction

$$(-1)^{2/6} = (-1)^{1/3} = -1$$
$$\text{but} \quad (-1)^{2/6} = ((-1)^2)^{1/6} = 1^{1/6} = 1.$$

91. Notice that by §82, §83 and §89 continuity of functions is preserved under scalar multiplication, modulus, sum, product, quotient, composition and inverse where the operations are possible. Deduce from a combination of these facts that functions like

$$(x^2+1)^{-1/2} \quad (x \varepsilon R) \quad \text{and} \quad (x^2-3x+8)^{2/3} \quad (x \varepsilon R)$$

are continuous on their domains.

<u>EXAMPLE</u>. Find $\lim_{x \to 0} (x^2-3x+8)^{2/3}$.

<u>Solution</u>. The function F defined by $F(x) = (x^2-3x+8)^{2/3}$ is continuous on R. So $\lim_{x \to 0} F(x) = F(0) = 8^{2/3} = 4.$

EXAMPLE.   Find $\lim\limits_{n\to\infty}\left(1+\dfrac{6}{n}\right)^{1/5}$.

Solution.   The required limit $= \lim\limits_{x\to 0} (1+x)^{1/5} = 1$ (by continuity).

92. TWO USEFUL LIMITS

RESULT.   Let $a > 0$.   Then, as $n \to \infty$,
$$n^{1/n} \to 1 \quad \text{and} \quad a^{1/n} \to 1.$$

Proof.   Use §61 with $x = n^{1/n}$ and $k = 2$.   This gives
$$(n^{1/n})^n > \binom{n}{2} (n^{1/n}-1)^2,$$
i.e. $n > \frac{1}{2}n(n-1)(n^{1/n}-1)^2$.

On rearrangement this gives
$$1 + \left(\frac{2}{n-1}\right)^{\frac{1}{2}} > n^{1/n} > 1,$$
whereupon a sandwich result shows that $n^{1/n} \to 1$ as $n \to \infty$.   The result for $a^{1/n}$ similarly.

(See also the more flexible general method of §211.)

EXAMPLES 3
(All numbers in these examples are real.)

1.   The function f is defined on R by
$$f(x) = \begin{cases} 4x+3 & (x > 0), \\ 0 & (x = 0), \\ -4x-3 & (x < 0). \end{cases}$$
Show that $\lim\limits_{x\to 0+} f(x)$ and $\lim\limits_{x\to 0-} f(x)$ both exist.   Does $\lim\limits_{x\to 0} f(x)$ exist?

2.   Let [·] denote the integral part function, i.e. [x] is the greatest integer less than or equal to x.   Prove that $\lim\limits_{x\to 1} [x]/x$ does not exist but that $\lim\limits_{x\to\infty} [x]/x = 1$.

3.   The function f is defined on R by
$$f(x) = \frac{x^k(5x^2+4)}{x^4-4x^2+6}, \text{ where } k\epsilon Z.$$ State without proof for

C

which values of $k \varepsilon Z$ the following situations occur as $x \to \infty$:

(a) $f(x) \to \infty$; (b) $f(x) \to 0$; (c) $f(x)$ tends to a non-zero limit.

4. Prove from the definition that the function f defined on $R-\{-6,3\}$ by

$$f(x) = \frac{x^2+9}{(x+6)(x-3)}$$

tends to infinity as $x \to 3+$.

5. Prove from the definition that f defined on $R-\{-1,6\}$ by

$$f(x) = \frac{(x-3)}{(x+1)(x-6)^2}$$

tends to infinity as $x \to 6$.

6. Prove from the definition of continuity that the function f defined on R by

$$f(x) = \begin{cases} 5x & (x \text{ rational}), \\ -3x & (x \text{ irrational}). \end{cases}$$

is continuous at $x = 0$ but is discontinuous at every other point of R.

7. State the values of $x \varepsilon R$ at which the integral part function (defined in question 2) is

(a) continuous on the right, (b) continuous on the left, (c) continuous.

8. The function f is defined on R by

$$f(x) = \begin{cases} 0 & (x \text{ rational}), \\ 1 & (x \text{ irrational}). \end{cases}$$

Prove that f is discontinuous at every point of R. Modify the definition of f to construct a function g which is discontinuous at every point of R but such that g^2 is continuous at every point of R.

9. The function f is defined on R by

$$f(x) = \begin{cases} 4x^2 & (x \text{ rational}), \\ 4 & (x \text{ irrational}). \end{cases}$$

Prove that f is continuous at $x = 1$ but that f is discontinuous at $x = 0$.

10. Show that the equation $x^3+3x^2-5x-3 = 0$ has a solution in each of the intervals $]-5,-4[,]-1,0[$ and $]1,2[$.

11.   Let p be a polynomial function of odd degree.
Prove that there exists at least one point c∈R such
that p(c) = 0.
   Give an example to show that this need not be true for
a polynomial of even degree.

12.   (i) Give an example of a continuous function
f: R → R for which there is no point c such that
f(c) = c.

   (ii) Give an example of a continuous function
f:]0,1[→]0,1[for which there is no point c such
that f(c) = c.

   (These examples illustrate the failure of the fixed
point result of §88 if the interval in question is not a
closed interval [a,b].)

13.   The function f: R → R is continuous and f
maps [a,b] into [c,d] and maps [c,d] into [a,b],
where [a,b] ∩ [c,d] = ∅.   By considering f∘f, prove
that there exist real numbers p and q such that
f(p) = q and f(q) = p.   Deduce that there exists x∈R
such that f(x) = x.   (See also the solution of Ex.5.15.)

14.   The function f is continuous, positive and
unbounded on R and inf f = 0.   Prove that the range
of f is]0,∞[.

15.   Let f be continuous and positive on R.
Suppose that f and 1/f have the same range.   Prove
that 1 is a member of the range of f.

16.   Let f be a continuous function on [a,b] such
that f(a) > 0 and f(b) < 0.   Show that the set S
defined by S = {x: x∈[a,b],f(x) = 0} contains its
infimum.

17.   Use the results of §92 together with the sandwich
theorem to evaluate the limits as n → ∞ of
$$n^{3/n}, \quad (2n+1)^{1/n}, \quad 5^{1/4^n}, \quad (3n)^{1/n^2}, \quad (n^2+n+1)^{1/n}.$$

18.   (i) Prove that if $\{a_n\}$ is a sequence such that
$$\frac{a_{n+1}}{L} = \left(\frac{a_n}{L}\right)^{1/4} \quad (n\in N)$$
where L ≠ 0, then
$$\frac{a_{n+1}}{L} = \left(\frac{a_1}{L}\right)^{1/4^n} \quad (n\in N).$$
Deduce that $a_n \to L$ as n → ∞.

(ii) Let $\alpha > 0$ and define a sequence by taking a_1 as any positive number and $a_{n+1} = (\alpha a_n)^{1/4}$ for all $n \varepsilon N$. Use the result of part (i) to show that $a_n \to \alpha^{1/3}$ as $n \to \infty$. (This gives a method whereby a calculator with a square root key can be used to find cube roots. This example was drawn to my attention by Professor R.A. Rankin. See also Ex.13.1.)

19. Give an example to show that if f is strictly increasing on a domain D, then f^2 can be strictly decreasing on D.

20. Prove that a rational real function cannot oscillate boundedly as $x \to \infty$.

21. Suppose f and g are defined on $]0,\infty[$ and that $f(x) \to L$ and g oscillates boundedly as $x \to 0+$. Prove by a contradiction argument that $f+g$ oscillates boundedly as $x \to 0+$.

22. Show that the function g defined on R by $g(x) = x-[x]$ is periodic on R. State the points at which g is continuous.

23. A function $f: R \to R$ is called <u>even</u> if $f(-x) = f(x)$ for all $x \varepsilon R$ (e.g. $f(x) = x^2$), and is called <u>odd</u> if $f(-x) = -f(x)$ for all $x \varepsilon R$ (e.g. $f(x) = x^3$).

For an arbitrary function $g: R \to R$, define g_1 and g_2 on R by
$$g_1(x) = g(x) + g(-x), \quad g_2(x) = g(x) - g(-x).$$
Show that g_1 is even and g_2 is odd and deduce that g is expressible as the sum of an even function and an odd function. Show that this representation is unique.

Deduce that an even polynomial function consists of even powers of x only, while an odd polynomial function consists of odd powers of x only.

24. For real functions f and g prove that if $f(x) \to A$ and $g(x) \to B$ as $x \to c+$, then $f(x)g(x) \to AB$.

25. Let f be an increasing function on R. Show that for every point $c \varepsilon R$, $\lim_{x \to c+} f(x)$ and $\lim_{x \to c-} f(x)$ both exist.

26. Prove that a real periodic function cannot tend to infinity as $x \to \infty$. Can such a function be unbounded?

27. The real function f is continuous on R and
$f(2x) = (f(x))^2$ for all xϵR. Prove that $f(0) = 0$
or 1. If it is given that $f(1) \neq 0$, prove that
$f(0) = 1$.

28. (An alternative characterisation of continuity)
Prove that the following two statements about the real
function f are equivalent:

(i) f is continuous at aϵR;

(ii) for all sequences $\{a_n\}$ such that $a_n \to a$, it
follows that $f(a_n) \to f(a)$.

29. Let f: R $\to$ R be an arbitrary function. Let
$f^+(x) = \max(f(x),0)$ and $f^-(x) = -\min(f(x),0)$. [The
functions f^+ and f^- are called the positive and
negative parts of f.] Show that f^+ and f^- are
non-negative functions and that $f = f^+ - f^-$. Deduce
that an arbitrary function can be expressed as the
difference of two non-negative functions. Demonstrate
by a particular example that this expression is not
unique.

IV. METRIC SPACES AND COMPACTNESS

---

93. METRIC

In R, $|x-y|$ measures the distance between x and y. Similarly in R^2, R^3 or more generally R^n $(n \geq 4)$, we can define a distance between two points. In R^3, for example

$$\{(x_1-y_1)^2 + (x_2-y_2)^2 + (x_3-y_3)^2\}^{\frac{1}{2}}$$

measures the distance between (x_1,x_2,x_3) and (y_1,y_2,y_3). A metric is effectively such a distance defined between <u>every</u> pair of points of a set S.

<u>Definition</u>. Let S be a non-empty set. Then a mapping $d: S \times S \to R$ is a <u>metric</u> provided that the following three conditions are satisfied:

(M1) $d(x,y) \geq 0$ for all $x,y \in S$, and $d(x,y) = 0$ if and only if $x = y$.

(M2) $d(x,y) = d(y,x)$ for all $x,y \in S$.

(M3) (<u>Triangle inequality</u>) $d(x,z) \leq d(x,y) + d(y,z)$ for all $x,y,z \in S$.

94. METRIC SPACE

A set S equipped with a metric d is a <u>metric space</u>. Examples of metric spaces are:

(i) R with $d(x,y) = |x-y|$.

(ii) $[0,1]$ with $d(x,y) = |x-y|$.

(iii) C with $d(z,w) = |z-w|$.

(iv) R^3 with $d(x,y) = \{\sum_{i=1}^{3} (x_i-y_i)^2\}^{\frac{1}{2}}$

where $x = (x_1,x_2,x_3)$ and $y = (y_1,y_2,y_3)$.

(v) R^n with $d(x,y) = \{\sum_{i=1}^{n} (x_i-y_i)^2\}^{\frac{1}{2}}$

where $x = (x_1,x_2,\ldots,x_n)$ and $y = (y_1,y_2,\ldots,y_n)$ and $n \in N$.

The metrics given here are the usual metrics for R, [0,1], C, R^3 and R^n and they are often implicitly understood on these sets if not otherwise stated.

95. CONVERGENCE IN A METRIC SPACE

The definition of this is similar to the corresponding definition for convergence of a sequence in R:

Definition. The sequence $\{x_n\}$ in a metric space (S,d) <u>tends to</u> aεS <u>as</u> $n \to \infty$ means:
$$\forall \varepsilon > 0, \ \exists X \ \text{such that} \ d(x_n,a) < \varepsilon \ \forall n > X.$$

Notice that the definition demands that x_n lies within distance ε of the point a whenever n lies beyond the threshold X.

<u>EXAMPLE</u>. In R^3 with the metric of §94(iv), the sequence $\{x_n\}$ defined by $x_n = ((8n+3)/n,(n-2)/n,(5n+6)/n)$ tends to (8,1,5) as $n \to \infty$.

<u>Solution</u>. $d(x_n,(8,1,5)) = \{\left(\frac{3}{n}\right)^2 + \left(\frac{2}{n}\right)^2 + \left(\frac{6}{n}\right)^2\}^{\frac{1}{2}} = \frac{7}{n}$
$$< \varepsilon \ \forall n > \frac{7}{\varepsilon}.$$

96.

Recall the neighbourhood $N(c,\delta) =]c-\delta,c+\delta[$ in R. To extend this idea to a general metric space (S,d), for a point cεS and for $\delta > 0$, define
$$N(c,\delta) = \{x: x\varepsilon S, d(x,c) < \delta\}$$
$$\bar{N}(c,\delta) = \{x: x\varepsilon S, d(x,c) \leq \delta\}.$$

$N(c,\delta)$ and $\bar{N}(c,\delta)$ are the <u>open</u> and <u>closed balls</u> of radius δ with centre c.

For R, R^2 and R^3 with the metrics of §94, the open balls are as follows. In R, $N(c,\delta)$ is the interval $]c-\delta,c+\delta[$. In R^2, $N(c,\delta)$ is the open disc with centre c and radius δ. In R^3, $N(c,\delta)$ is the open solid sphere with centre c and radius δ.

The fact that $x_n \to a$ as $n \to \infty$ means that for all open balls $N(a,\varepsilon)$ of positive radius, there exists a threshold X such that x_n lies inside $N(a,\varepsilon)$ for all n beyond X.

97. Different metrics can be defined on the same space.

For example look at R^2. For $x = (x_1, x_2)$ and $y = (y_1, y_2)$, take

$$d_1(x,y) = \{(x_1-y_1)^2 + (x_2-y_2)^2\}^{\frac{1}{2}},$$

$$d_2(x,y) = \max(|x_1-y_1|, |x_2-y_2|),$$

$$d_3(x,y) = \begin{cases} 0 & \text{if } x = y, \\ 1 & \text{if } x \neq y. \end{cases}$$

For these metrics on R^2 the open ball $N((2,2),1)$, which has centre $(2,2)$ and radius 1, has the following shapes

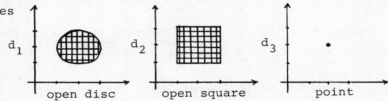

98. <u>EXAMPLE</u>. In R^2 consider the sequence $\{x_n\}$, defined by $x_n = (2+3/n, 2+4/n)$, under the various metrics in §97.

<u>Solution</u>. $\{x_n\}$ tends to $(2,2)$ under d_1 and d_2 but <u>not</u> under d_3. To see this, notice that:

$$d_1(x_n, (2,2)) = 5/n < \varepsilon \qquad \forall n > 5/\varepsilon,$$

$$d_2(x_n, (2,2)) = 4/n < \varepsilon \qquad \forall n > 4/\varepsilon,$$

$$d_3(x_n, (2,2)) = 1 \text{ for all } n \geq 1, \text{ and it is}$$
never less than ε if for example $\varepsilon = \frac{1}{2}$.

Study the geometrical significance of these results in relation to the shapes of the open balls in §97.

99. EQUIVALENT METRICS

Two metrics on the same set S are <u>equivalent</u> if the sequences convergent with respect to one are also convergent with respect to the other and vice versa. Of the three metrics on R^2 in §97, only d_1 and d_2 are equivalent.

The following result shows how to decide whether two metrics are equivalent or not by examining the shapes of the open balls they determine.

RESULT. Let d_1 and d_2 be metrics on a set S. Then

$\qquad$ (α) d_1 and d_2 are equivalent

$\Longleftrightarrow$ (β) every d_1-open ball contains a concentric d_2-open ball and every d_2-open ball contains a concentric d_1-open ball.

Proof. (To prove (β) $\Longrightarrow$ (α)) Suppose (β) holds. Take a sequence $\{x_n\}$ such that $x_n \to a$ in (S,d_1) as $n \to \infty$. Choose $\varepsilon > 0$. [We want to show that $x_n \to a$ in (S,d_2).] From hypothesis (β) there exists $\delta > 0$ such that $N_1(a,\delta) \subseteq N_2(a,\varepsilon)$. Also, since $x_n \to a$ in (S,d_1), $x_n \in N_1(a,\delta)$ for all $n > X$. So $x_n \in N_2(a,\varepsilon)$ for all $n > X$, i.e. $x_n \to a$ in (S,d_2). So a d_1-convergent sequence is d_2-convergent. By a similar argument every d_2-convergent sequence is d_1-convergent. So d_1 and d_2 are equivalent. So we conclude that (β) $\Longrightarrow$ (α).

$\qquad$ (To prove (α) $\Longrightarrow$ (β)) We shall prove (equivalently) that $\neg(\beta) \Longrightarrow \neg(\alpha)$. Suppose therefore that the negation of (β) is true. So without loss of generality assume that

$\neg(\; \forall a \varepsilon S, \; \forall \varepsilon > 0, \; \exists \delta > 0 \quad \text{s.t.} \quad N_1(a,\delta) \subseteq N_2(a,\varepsilon))$,

i.e. $\exists a \varepsilon S$ & $\exists \varepsilon > 0$, s.t. $\forall \delta > 0$, $N_1(a,\delta) \not\subseteq N_2(a,\varepsilon)$.

For this a and for this ε, let δ successively take the values $1, 1/2, 1/3, \dots$ and hence choose a sequence of points $x_1, x_2, x_3, \dots$ such that

$$x_n \in N_1(a, \tfrac{1}{n}) \qquad \text{but} \qquad x_n \notin N_2(a,\varepsilon).$$

Then $d_1(x_n,a) < 1/n$ but $d_2(x_n,a) \geq \varepsilon$. So $x_n \to a$ in (S,d_1) but $x_n \not\to a$ in (S,d_2). So d_1 and d_2 are not equivalent. So $\neg(\beta) \Longrightarrow \neg(\alpha)$ as required.

$\qquad$ Notice how this result applied to the metrics of §97 gives the equivalence of d_1 and d_2 (because we can place a square inside every disc and a disc inside every square) and gives the inequivalence of d_1 and d_3 (because we cannot place a disc inside a point).

EXAMPLE. For points $x = (x_1, x_2)$ and $y = (y_1, y_2)$ in R^2, define $d(x,y) = |x_1 - y_1| + |x_2 - y_2|$. Show that the ball $N((0,0), \delta)$ is a square with vertices at $(\delta, 0)$, $(0, \delta)$, $(-\delta, 0)$, $(0, -\delta)$. Deduce that d is equivalent to the metrics d_1 and d_2 of §97.

100. THE DISCRETE METRIC

Any set can be made into a metric space with the metric defined by

$$d(x,y) = \begin{cases} 0 & \text{if} \quad x = y, \\ 1 & \text{if} \quad x \neq y. \end{cases}$$

This is called the <u>discrete metric</u> on the set. A case in point is the metric d_3 on R^2 in §97.

For the <u>discrete metric on R</u>, notice:-

(i) $N(5, \tfrac{1}{2}) = \{5\}$, $N(5,1) = \{5\}$, $N(5,2) = R$. (Recall that $N(a, \delta) = \{x: d(x,a) < \delta\}$.)

(ii)   The sequence $\{x_n\}$ converges to 5 if and only if $\exists X$ such that $\forall n > X$, $x_n = 5$. So, for example, the sequence $\{x_n\}$, where $x_n = 5 + 1/n$ ($n \epsilon N$) does <u>not</u> converge to 5 in this metric.

101. EXAMPLE.
(i) Show that (R^2, d) is a metric space, where d is defined by

$$d((x_1, x_2), (y_1, y_2)) = \begin{cases} |x_2 - y_2| & \text{if} \quad x_1 = y_1, \\ |x_1 - y_1| + |x_2| + |y_2| & \text{if} \quad x_1 \neq y_1. \end{cases}$$

(This metric is effectively the distance between two points in a library of shelves parallel to the y-axis and with a connecting corridor along the x-axis only.)

(ii) Draw the neighbourhoods $N((2,3),1)$ and $N((2,0),1)$ in this space. Show that the sequence $\{(2, 3 + \tfrac{1}{n})\}$ ($n \epsilon N$) converges to $(2,3)$ in (R^2, d) while the sequence $\{(2 + \tfrac{1}{n}, 3)\}$ ($n \epsilon N$) does <u>not</u> converge to $(2,3)$ in (R^2, d).

(iii) Is d equivalent to the metric d_2 for R^2 in §97?

<u>Solution</u>.   (i) Write x for (x_1,x_2) etc.   It is
clear that $d(x,y) \geq 0$, that $d(x,y) = 0$ if and only if
$x = y$ and that $d(x,y) = d(y,x)$.   So (M1) and (M2) are
true.   The triangle inequality (M3), i.e.
$d(x,z) \leq d(x,y) + d(y,z)$, is fairly clear from the library
analogy but to prove it analytically we split the work
into cases.

$\quad$ <u>Case 1</u>.   If $x_1 = z_1$,
$d(x,z) = |x_2-z_2| \leq |x_2-y_2| + |y_2-z_2| \leq d(x,y) + d(y,z)$.

$\quad$ <u>Case 2</u>.   If $x_1 \neq z_1$ and $y_1 = x_1$ (or similarly
$x_1 \neq z_1$ and $y_1 = z_1$),
$$\begin{aligned}
d(x,z) &= |x_1-z_1| + |x_2| + |z_2| \\
&= |y_1-z_1| + |(x_2-y_2) + y_2| + |z_2| \\
&\leq |y_1-z_1| + |x_2-y_2| + |y_2| + |z_2| \\
&= |x_2-y_2| + (|y_1-z_1| + |y_2| + |z_2|) \\
&= d(x,y) + d(y,z).
\end{aligned}$$

$\quad$ <u>Case 3</u>.   If x_1, y_1 and z_1 are all different,
$$\begin{aligned}
d(x,z) &= |x_1-z_1| + |x_2| + |z_2| \\
&= |(x_1-y_1) + (y_1-z_1)| + |x_2| + |z_2| \\
&\leq |x_1-y_1| + |y_1-z_1| + |x_2| + |z_2| \\
&\leq (|x_1-y_1| + |x_2| + |y_2|) + (|y_1-z_1| + |y_2| + |z_2|) \\
&= d(x,y) + d(y,z).
\end{aligned}$$

So (M3) is proved, and (R^2,d) is a metric space.

(ii)

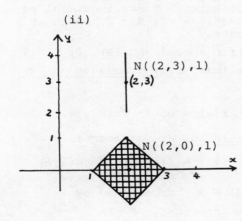

$N((2,3),1)$ is a
straight line seg-
ment, the points on
it being the only
points in R^2 with-
in d-distance 1 of
$(2,3)$.

$N((2,0),1)$ is an
open square as shown.

Let $x_n = (2, 3 + \frac{1}{n})$, $y_n = (2 + \frac{1}{n}, 3)$ and let $a = (2,3)$. Then

$$d(x_n, a) = |3 + \frac{1}{n} - 3| = \frac{1}{n} \to 0 \quad \text{as} \quad n \to \infty.$$

So $x_n \to a$ as $n \to \infty$.

Also, $d(y_n, a) = |(2 + \frac{1}{n}) - 2| + 3 + 3$

$$= (6 + \frac{1}{n}) \not\to 0 \quad \text{as} \quad n \to \infty.$$

So $y_n \not\to a$ as $n \to \infty$.

(iii) For the metric d_2 of §97,

$$d_2(y_n, a) = \max(|2 + \frac{1}{n} - 2|, |3-3|)$$

$$= \frac{1}{n} \to 0 \quad \text{as} \quad n \to \infty.$$

So $y_n \to a$ in (R^2, d_2). But from above, $y_n \not\to a$ in (R^2, d). So d and d_2 are <u>not</u> equivalent metrics on R^2.

(As further evidence of the inequivalence of d and d_2, notice that it is impossible to find a d_2-open ball inside the d-open ball with centre $(2,3)$ and radius 1, i.e. it is impossible to place a square inside the line segment drawn in part (ii).)

102. CONTINUITY IN A GENERAL METRIC SPACE

For a function $f: R \to R$, the idea of the limit of f as x tends to a and the idea of continuity of f at a have already been dealt with. These ideas can be generalised by considering functions $f: S \to T$, where S and T are general metric spaces.

Let S and T be metric spaces and let $f: S \to T$ be a mapping. The statement that <u>f(x) tends to b as x tends to a</u> means:

$\forall \varepsilon > 0$, $\exists \delta > 0$ such that $f(x) \in N(b, \varepsilon)$ $\forall x \in N(a, \delta)$.

The statement that <u>f is continuous at a</u> means:

$\forall \varepsilon > 0$, $\exists \delta > 0$ such that $f(x) \in N(f(a), \varepsilon)$ $\forall x \in N(a, \delta)$.

<u>EXAMPLE.</u> The function $f: R^3 \to R$ is defined by $f(x_1, x_2, x_3) = 4(x_1^2 + x_2^2 + x_3^2)^2$. Prove that f is continuous at $(0,0,0)$.

__Solution__.   Choose $\varepsilon > 0$.   Write $x = (x_1, x_2, x_3)$ and
$0 = (0,0,0)$.   Then
$$|f(x) - f(0)| = 4(x_1^2 + x_2^2 + x_3^2)^2 < \varepsilon \quad \text{for all} \quad x \in N(0, \frac{\varepsilon^{\frac{1}{4}}}{\sqrt{2}}).$$

A very useful criterion for continuity (which looks at the behaviour of sequences) is given in Ex.3.28.

103.   OPEN SETS

__Definition__.   A subset A of a metric space is __open__ if for every point $a \varepsilon A$, there exists a number $\varepsilon > 0$ such that $N(a, \varepsilon) \subseteq A$.

Notice: (i) ε may vary with a.
(ii) In R^2 with the metric d_1 of §97, the triangular region $A = \{(x,y): 0 < y < x < 1\}$ is an open set, since round every point we can draw a small open ball (i.e. a disc) contained wholly within A.
On the other hand the set $B = A \cup \{(0,0)\}$, i.e. the triangular region together with the origin, is __not__ open in (R^2, d_1) because no open ball $N((0,0), \varepsilon)$ lies wholly within B.
(iii) $N(a, \varepsilon)$ is an open set itself.
(iv) For R with the discrete metric every point is an open set.   For $\{5\} = N(5, \frac{1}{2}) \subseteq \{5\}$.

__Definition__.   A subset A of a metric space is __closed__ if the complement of A is open.

__EXAMPLE__.   In R^2 with the metric d_1 of §97, the sets $A = \{(x,y): x^2 + y^2 \leq 1\}$ and $B = \{(x,y): x^2 + y^2 = 1\}$ are both closed.

__EXAMPLE__.   Show that if d_1 and d_2 are equivalent metrics on a set S then they determine the same open sets.

104.   COMPACTNESS

A family of sets, $\{C_\alpha: \alpha \varepsilon I\}$ (where I is a set of suffices), is a __covering__ of a set A if $A \subseteq \bigcup_{\alpha \varepsilon I} C_\alpha$.

An <u>open covering</u> is a covering by open sets.

<u>Definition</u>. A subset A of a metric space is <u>compact</u> if every open covering of A contains a finite sub-covering (i.e. a <u>finite</u> number of the sets of the original covering form a covering).

<u>EXAMPLES</u>. (i) In R with the usual metric every closed interval [a,b] is compact. This is not obvious but is proved in §105. Every finite subset of R is compact. This is obvious.

(ii) In R with the discrete metric the only com-pact sets are the finite sets. For, by §100(i), every point is an open set itself, so that every infinite set has an open covering from which no finite subcovering can be selected.

(iii) In R^n $(n \geq 1)$ a set is compact if and only if it is closed and bounded. This is not obvious.

105. EVERY BOUNDED CLOSED INTERVAL ON THE REAL LINE IS COMPACT

The following far-reaching result applies to R with the usual metric. This result in one form or another is called the Borel covering theorem or the Heine-Borel theorem.

<u>RESULT</u>. Let $a, b \in R$ with $a < b$. Then [a,b] is compact.

<u>Proof</u>. Let $\{V_\alpha : \alpha \in I\}$ be an open covering of [a,b]. Define the set C by the following rule:

$c \in C \iff a < c \leq b$ <u>and</u> [a,c] can be covered by finitely many of the sets V_α.

Now there exists $\alpha \in I$ such that $a \in V_\alpha$. Then since V_α is open, there exists $k > 0$ such that $]a-k, a+k[\subseteq V_\alpha$. So $[a, a+\tfrac{1}{2}k]$ is covered by finitely many (in fact by just one) of the sets V_α $(\alpha \in I)$. So $(a+\tfrac{1}{2}k) \in C$. So $C \neq \emptyset$ and C is bounded above by b. So C has supremum σ where $a < \sigma \leq b$.

Now suppose that $\sigma \neq b$. So $\sigma < b$. Then since $\sigma \in V_\beta$ for some β and since $\sigma < b$, there exists an interval $[\sigma-\varepsilon, \sigma+\varepsilon]$ with $\varepsilon > 0$ and contained in V_β and such that $a < \sigma-\varepsilon < \sigma+\varepsilon < b$. Now however, $[a, \sigma-\varepsilon]$ can be covered by finitely many sets V_α (by the definition of σ) and V_β covers $[\sigma-\varepsilon, \sigma+\varepsilon]$. So

$[a,\sigma+\varepsilon]$ can be covered by the finite number of sets covering $[a,\sigma-\varepsilon]$ together with V_β, i.e. $[a,\sigma+\varepsilon]$ can be covered by finitely many sets V_α. This contradicts the definition of σ. So our assumption above is false and $\sigma = b$. (This shows that $b = \sup C$ but does not guarantee that $b\varepsilon C$.)

Also, however, there exists a set V_γ such that $b\varepsilon V_\gamma$ and $]b-\delta,b] \subseteq V_\gamma$ with $a < b-\delta$. Then $[a,b-\delta]$ can be covered by finitely many sets V_α (since $b = \sup C$) and $]b-\delta,b]$ is covered by V_γ. So $[a,b]$ can be covered by the finite number of sets covering $[a,b-\delta]$ together with V_γ, i.e. $[a,b]$ can be covered by finitely many sets V_α. So $[a,b]$ is compact.

This result is a particular case of the important result in R^n ($n \geq 1$): <u>A subset of R^n is compact if and only if it is closed and bounded.</u>

CONTINUOUS FUNCTIONS ON COMPACT SETS

106. On a <u>compact</u> set in a metric space a continuous function (i) is bounded, (ii) attains both its supremum and infimum, and (iii) is uniformly continuous. In §107-§112 we examine these results and their consequences in the case when the metric space is R and the compact set is a closed interval.

107. BOUNDEDNESS AND ATTAINMENT OF SUPREMUM AND INFIMUM

RESULT. Let $f: [a,b] \to R$ be a continuous function. Then, (i) f is bounded on $[a,b]$, and (ii) f attains its supremum and infimum on $[a,b]$.

Proof. (i) For all $c\varepsilon[a,b]$, f is continuous at c (with modifications at endpoints). So, taking $\varepsilon = 1$, deduce that

$\exists \delta_c > 0$ such that $|f(x)-f(c)| < 1$ $\forall x\varepsilon N(c,\delta_c) \cap [a,b]$.

So $\forall x\varepsilon N(c,\delta_c) \cap [a,b]$,

$$|f(x)| \leq |f(x)-f(c)| + |f(c)| \quad \text{(triangle inequality)}$$
$$\leq 1 + |f(c)| = M(c), \text{ say.}$$

The open sets $N(c,\delta_c)$ (for all $c\epsilon[a,b]$) form an open covering of $[a,b]$. But $[a,b]$ is compact. So there exists a finite subcovering by the open sets centred at $c_1,c_2,\ldots,c_k$. Then for all $x\epsilon[a,b]$,

$$|f(x)| \le \max(M(c_1),M(c_2),\ldots,M(c_k)),$$

i.e. f is bounded on $[a,b]$.

(ii) Suppose that f does <u>not</u> attain its supremum M. Then $g: [a,b] \to R$ defined by $g(x) = 1/(M-f(x))$ is positive and continuous on $[a,b]$, and so by part (i) g is bounded above by a positive number K. So $1/(M-f(x)) < K$ for all $x\epsilon[a,b]$. So $f(x) < M-1/K$ for all $x\epsilon[a,b]$, which contradicts the fact that $M = \sup f$. So f does attain its supremum. Similarly for infimum.

To generalise this result consider a continuous function $f: S \to R$, where S is a compact subset of a general metric space. Then with minor modifications the above proof will show that f is bounded on S and attains its bounds.
In design theory in statistics, the supremum of a continuous function $f: R^n \to R$ on a compact subset A of R^n is often at issue. A result similar to the above assures us that there <u>does</u> exist a point of A at which sup f is attained.

108. In the shade of §107 notice that a continuous function on an open or half-open interval need not be bounded and even if it is bounded it need not attain its bounds.

<u>EXAMPLES</u>. (i) Define $f:]0,1[\to R$ by $f(x) = 1/x$. Here f is continuous but is unbounded above and fails to attain its infimum.

(ii) Define $f:]0,1[\to R$ by $f(x) = 3x$. Here f is continuous and bounded but attains neither its supremum nor its infimum.

(iii) Define $f: [0,1] \to R$ by $f(0) = f(1) = 5$, $f(x) = 1/x$ $(x \ne 0,1)$. Here f is defined on a closed interval but is not continuous. f is unbounded above and fails to attain its infimum.

109. IMAGE OF A CLOSED INTERVAL UNDER A CONTINUOUS FUNCTION

By §107 the image of a closed interval under a continuous function is a bounded set and moreover the image contains its supremum σ and its infimum τ. Also by the intermediate value theorem of §87, the image must contain all points between τ and σ. So the image is the closed interval $[\tau,\sigma]$. (If $\tau = \sigma$ the image is a one point set.)

This result is evidence of the fact that compactness and connectedness are preserved under the action of continuous functions.

110. UNIFORM CONTINUITY

Definition. That the real function f is uniformly continuous on the interval I means:

$$\forall\, \varepsilon > 0, \quad \exists \delta > 0 \quad \text{such that}$$
$$|f(x)-f(y)| < \varepsilon \quad \forall x,y\varepsilon I \quad \text{with} \quad |x-y| < \delta.$$

N.B. Here δ is independent of the position of x and y within the interval I (except that they must be less than δ apart). Asking for f to be uniformly continuous on I is asking for more than for f to be continuous on I. The following example and the valuable simple result of §136 give insight into the meaning of uniform continuity on an interval.

EXAMPLE. Let f and g be the real functions defined on $]0,1[$ by $f(x) = 15x^2$ and $g(x) = 1/x$. Prove that
(i) f is uniformly continuous on $]0,1[$, but that
(ii) g is not uniformly continuous on $]0,1[$.

Solution. (i) $|f(x)-f(y)| = 15|x-y||x+y|$
$$\leq 30|x-y| \quad \forall x,y\varepsilon]0,1[.$$
To get $30|x-y| < \varepsilon$, take $\delta = \varepsilon/30$. So $\forall\, \varepsilon > 0$, $\exists \delta (= \varepsilon/30)$ such that $|f(x)-f(y)| < \varepsilon$ whenever $x,y\varepsilon]0,1[$ and $|x-y| < \delta$. So f is uniformly continuous on $]0,1[$.

(ii) Suppose that g is uniformly continuous on $]0,1[$. Take $\varepsilon = \frac{1}{2}$ in the definition and find that $\exists\, \delta > 0$ such that $|g(x) - g(y)| < \frac{1}{2} \,\forall x,y\varepsilon]0,1[$ such that $|x-y| < \delta$. In $]0,\delta[$ choose points $x = 1/(2N)$

and $y = 1/N$ where $N \varepsilon N$. Then
$$|g(x)-g(y)| = |2N-N| = N > \tfrac{1}{2},$$
even although $|x-y| < \delta$. This is a contradiction.
So g is not uniformly continuous on $]0,1[$.

Notice that it is the existence of points x,y as close together as we like and such that $|g(x)-g(y)|$ is large which ensures that g cannot be uniformly continuous on $]0,1[$.
Notice also that the statement "f is uniformly continuous", in which no interval is specified, is meaningless.

111. On a <u>non-closed</u> interval, as in the example of §110, a continuous function may or may not be uniformly continuous. On a <u>closed</u> interval however the situation is not in doubt.

RESULT. Let f be continuous on the closed interval $[a,b]$. Then f is uniformly continuous on $[a,b]$.

Proof. For each $c \varepsilon [a,b]$, find by the continuity of f, an open interval $I(c) =]c-\delta(c),c+\delta(c)[$ such that for all $x \varepsilon I(c) \cap [a,b]$, $|f(x)-f(c)| < \varepsilon/4$. So for each pair of points $x,y \varepsilon I(c) \cap [a,b]$, $|f(x)-f(y)| < \tfrac{1}{2}\varepsilon$, as in §48(iii).

The open sets $\{I(c): c \varepsilon [a,b]\}$ form an open covering of $[a,b]$ and since $[a,b]$ is compact there exists a finite subcovering by $I(c_1),I(c_2),\ldots,I(c_k)$. Let $\delta = \min(\delta(c_1),\delta(c_2),\ldots,\delta(c_k))$. Notice that each of $I(c_1),I(c_2),\ldots,I(c_k)$ has length $\geq 2\delta$.

Now take $x,y \varepsilon [a,b]$ with $|x-y| < \delta$ and $x < y$. Then <u>either</u> $[x,y]$ is straddled by an interval $I(c_r)$, in which case $|f(x)-f(y)| < \tfrac{1}{2}\varepsilon < \varepsilon$, <u>or</u> there are two overlapping intervals $I(c_s)$ and $I(c_t)$, one containing x, the other containing y and with a common point p. In this case
$$|f(x)-f(y)| \leq |f(x)-f(p)| + |f(p)-f(y)|$$
$$< \tfrac{1}{2}\varepsilon + \tfrac{1}{2}\varepsilon = \varepsilon.$$
So, for all pairs of points $x,y \varepsilon [a,b]$ with $|x-y| < \delta$, $|f(x)-f(y)| < \varepsilon$.

112. To generalise the definition of uniform continuity, let (S,d) be a metric space and consider a function $f: S \to R$. That f is <u>uniformly continuous on a subset A of S</u> means:

$$\forall \varepsilon > 0, \ \exists \delta > 0 \quad \text{such that}$$
$$|f(x)-f(y)| < \varepsilon \ \forall x,y \varepsilon A \quad \text{with} \quad d(x,y) < \delta.$$

A proof analogous to that in §111 shows that a real function continuous on a compact subset of a metric space is uniformly continuous on that subset.

LIMIT POINTS AND CAUCHY SEQUENCES

113. LIMIT POINTS

<u>Definition</u>. Let A be a set in a metric space and let p be a point (not necessarily in A). Then p is called a <u>limit point of A</u> if every open neighbourhood of p contains a point of A that is not p itself.

Look at these examples in R with its usual metric:

(a) $[0,1]$ has $0,1,\frac{1}{2},0\cdot363,...$ as limit points.

(b) $]0,1[$ has $0,1,\frac{1}{2},0\cdot363,...$ as limit points.

(c) $\{1,1/2,1/3,1/4,...\}$ has only 0 as a limit point.

<u>EXAMPLE</u>. Show that in R with the usual metric every real number is a limit point of the set Q of all rational numbers.

<u>EXAMPLE</u>. Show that if a subset A of R lies within the closed interval $[c,d]$ then no point outside $[c,d]$ is a limit point of A. (For such a point has an open neighbourhood which fails to intersect A.) Generalise this statement to general metric spaces.

114. BOLZANO-WEIERSTRASS THEOREM

<u>RESULT</u>. An infinite bounded subset A of R has at least one limit point.

<u>Proof</u>. Since A is bounded there is a closed interval $[c,d]$ such that $A \subseteq [c,d]$. Notice from the second example of §113 that all possible limit points lie in $[c,d]$.

Suppose now that A has no limit point. So no point in $[c,d]$ is a limit point. So for every $p\varepsilon[c,d]$, there exists an open set $N(p)$ such that $N(p)$

contains at most one point of A.   (The one possible
point would be p.)   So for each pϵ[c,d], there exists
an open set N(p) containing at most one point of A.
The sets   N(p) (pϵ[c,d])   form an open covering of the
compact set [c,d].   So there is a finite subcovering
by N(p$_1$),N(p$_2$),...,N(p$_k$).   So

$$A \subseteq [c,d] \subseteq \bigcup_{i=1}^{k} N(p_i).$$

But each N(p$_i$) contains at most one point of A.   So
A is a finite set.   This is a contradiction and so the
initial assumption is wrong.   So A has a limit point.

Notice that the proof modifies directly to a
general metric space.

EXAMPLE.   Use the Bolzano-Weierstrass theorem to prove
that a bounded real sequence has a convergent subsequence.
(Compare §118.)

115. CAUCHY SEQUENCES

Definition.   A sequence {a$_n$} of real or complex
numbers is called a Cauchy sequence if

$$\forall \epsilon > 0, \ \exists X \text{ such that } |a_m - a_n| < \epsilon \ \forall m,n > X.$$

By analogy, in a general metric space (S,d),
define a sequence of points {a$_n$} to be a Cauchy
sequence if

$$\forall \epsilon > 0, \ \exists X \text{ such that } d(a_m, a_n) < \epsilon \ \forall m,n > X.$$

In R, C and indeed in every metric space, a convergent sequence is automatically Cauchy.   To see this
for sequences in R and C, just replace ϵ by $\frac{1}{2}\epsilon$ in
the result of §48(iii).   For a general metric space the
argument is analogous.
The converse situation is much less simple.   In
some metric spaces (including R and C), every Cauchy
sequence is automatically convergent; such spaces are
called complete.   In others, a Cauchy sequence need not
be convergent.   The positive results for R and C
are proved in §116 and §117, while Ex.4.4 shows how a
Cauchy sequence can fail to converge in a non-complete
space.

116. CAUCHY SEQUENCES IN R

RESULT. A Cauchy sequence in R is bounded.

Proof. Let $\{a_n\}$ be a Cauchy sequence. Find $k \epsilon N$ such that $|a_r - a_s| < 1$ for all $r, s \geq k$. So for every $r \geq k$, $|a_r - a_k| < 1$. So, for every $r \geq k$,

$$|a_r| \leq |a_r - a_k| + |a_k| < 1 + |a_k|.$$

It follows that, for all $n \geq 1$,

$$|a_n| \leq \max(|a_1|, |a_2|, \ldots, |a_{k-1}|, 1 + |a_k|).$$

So $\{a_n\}$ is bounded.

RESULT. A Cauchy sequence in R is convergent.

Proof. From the previous result, $\{a_n\}$ is a bounded sequence and so by the example of §114 there exists a subsequence of $\{a_n\}$ tending to L, say.
  Choose $\varepsilon > 0$. Then

$$\exists X \quad \text{such that} \quad |a_m - a_n| < \tfrac{1}{2}\varepsilon \quad \forall n > X.$$

Choose $\{a_s\}$ a member of the convergent subsequence such that

$$|a_s - L| < \tfrac{1}{2}\varepsilon \quad \underline{\text{and}} \quad s > X.$$

Then, for every $n > X$,

$$|a_n - L| \leq |a_n - a_s| + |a_s - L| < \tfrac{1}{2}\varepsilon + \tfrac{1}{2}\varepsilon = \varepsilon.$$

So $a_n \to L$ as $n \to \infty$.

117. CAUCHY SEQUENCES IN C

  The ideas of boundedness and convergence for complex sequences are discussed in §164.

RESULT. A Cauchy sequence in C is bounded and convergent.

Proof. Let $\{a_n\}$ be the Cauchy sequence. Let $a_n = b_n + i c_n$ where $b_n, c_n \epsilon R$. Then since $|\text{Re } z| \leq |z|$ and $|\text{Im } z| \leq |z|$, notice that

$$|b_m - b_n| \leq |a_m - a_n| \quad \text{and} \quad |c_m - c_n| \leq |a_m - a_n|.$$

So $\{b_n\}$ and $\{c_n\}$ are <u>real</u> Cauchy sequences to which §116 applies. So $\{b_n\}$ and $\{c_n\}$ are bounded and convergent. Hence $\{a_n\}$ is bounded and convergent by §164.

118. OTHER DEFINITIONS OF COMPACTNESS

There are characterisations of compactness in a metric space other than the definition of §104. For example, a subset A of a metric space is <u>compact</u> if and only if every sequence of points in A has a subsequence convergent to a point of A.

EXAMPLES 4

1. Show that the following spaces are <u>not</u> metric spaces:
   (a) R with $d(x,y) = |x| + |y|$;
   (b) R with $d(x,y) = |x-y|^2$;
   (c) $[-2,2]$ with $d(x,y) = |x^2-y^2|$;
   (d) R with $d(x,y) = 2^{|x-y|}$.

2. Show that (R^2,d) where $d((x_1,x_2),(y_1,y_2)) = |x_1-y_1| + |x_2-y_2|$ is a metric space. Draw the open ball with centre $(0,0)$ and radius 1. Show that this metric is equivalent to d_2 in §97.

3. Let $S = [1,5]$ and let $d_1(x,y) = |x-y|$ and $d_2(x,y) = |x^2-y^2|$ for all $x,y \varepsilon S$. Verify that d_1 and d_2 are metrics on S.
   Prove that $2|x-y| \le |x^2-y^2| \le 10|x-y|$ for all $x,y \varepsilon S$ and deduce that d_1 and d_2 are equivalent metrics on S.

4. Let $S =]0,\infty[$ and let $d(x,y) = \left|\frac{1}{x} - \frac{1}{y}\right|$ for all $x,y \varepsilon S$. Prove that d is a metric on S. Prove that $\{1,2,3,4,\ldots\}$ is a Cauchy sequence in (S,d) and deduce that (S,d) is not complete.

5. Show that if d_1 and d_2 are metrics on a space S, then so also are e and f where

$$e(x,y) = d_1(x,y) + d_2(x,y),$$
$$f(x,y) = \max(d_1(x,y), d_2(x,y)).$$

6. Define $d: C \times C \to [0,\infty[$ by

$$d(z,w) = \begin{cases} |z-w| & \text{if } \arg z = \arg w, \\ |z|+|w| & \text{if } \arg z \neq \arg w. \end{cases}$$

(Take $\arg 0 = 0$.) It is <u>given</u> that d is a metric on C. (Notice that d is effectively the distance between points in a roundhouse railway engine shed, in which engines are parked on lines radiating out from a central turntable, which is the only means of moving between them.) Draw the neighbourhoods $N(0,1)$, $N(1,1)$, $N(1+i,1)$. Does the sequence $\{x_n\}$ tend to $1+i$ as $n \to \infty$ with respect to this metric in the cases when

(a) $x_n = (1+\frac{1}{n})+i(1+\frac{1}{n})$,   (b) $x_n = (1+\frac{1}{n})+i(1-\frac{1}{n})$,

(c) $x_n = 1+i(1-\frac{1}{n})$?

Is this metric equivalent to the usual metric on C?

7. In a metric space (S,d), prove that
$$|d(x,z)-d(y,z)| \leq d(x,y)$$
for all $x,y,z \varepsilon S$.

8. State which of the following subsets of R are closed (in the usual metric):

(a) $]1,3]$,   (b) $[0,1[\cup]1,2]$,   (c) Z,   (d) $\{1/n: n\varepsilon N\}$,

(e) $[3,4]\cup\{5\}$,   (f) Q,   (g) $\{0\}\cup\{1/n: n\varepsilon N\}$.

9. State the limit points (if any) of the following subsets of R (with the usual metric):

(a) $[1,3]\cup\{4\}$,   (b) $]1,3[$,   (c) Q,

(d) $\{1/n: n\varepsilon N\}$,   (e) Z.

10. Give an example of an open covering of $]0,1]$ with no finite subcovering.

11. The function f is both positive and continuous on R. Also $f(x) \to 0$ as $x \to \pm\infty$. Prove that f is bounded and attains its supremum. Give an example of a rational function with these properties.

12. Show that if f and g are uniformly continuous on an interval I then so also is $f+g$.

13. The function f is defined on [-4,4] by
$f(x) = 2x^2+5x+2$.   Prove from first principles (i.e. by
looking at $|f(x)-f(y)|$) that f is uniformly con-
tinuous on [-4,4].   (Actually the result of §111
guarantees the uniform continuity because f is <u>con-
tinuous</u> on the <u>closed</u> interval.)

14. Of the following two statements (a) and (b), one
says that f is <u>continuous</u> on R, the other says that f
is <u>uniformly continuous</u> on R.   Which is which?

   (a) $\forall \varepsilon > 0$, $\exists \delta > 0$ such that $\forall c \varepsilon R$,
                $\forall x \varepsilon N(c,\delta)$ $|f(x)-f(c)| < \varepsilon$.

   (b) $\forall \varepsilon > 0$, $\forall c \varepsilon R$, $\exists \delta > 0$ such that
                $\forall x \varepsilon N(c,\delta)$ $|f(x)-f(c)| < \varepsilon$.

15. The function f is uniformly continuous on R,
so that if we take $\varepsilon=1$ in the definition, we conclude
that there exists $\delta > 0$ such that
        $|f(x)-f(y)| < 1$ whenever $|x-y| \leq \delta$.
Get an estimate of $|f(n\delta)-f(0)|$ and hence show that
there exist constants A and B such that
            $|f(x)| \leq A+B|x|$ for all $x \varepsilon R$.

16. Use the result of Ex.15 to show that x^2 is <u>not</u>
uniformly continuous on R.
   Give an example to show that the product of two
functions that are uniformly continuous on R need <u>not</u>
be uniformly continuous on R.   (Contrast Ex.12.)

17. For each of the following functions, state
whether it is uniformly continuous on $]0,\infty[$:
   (a) x, (b) x^3, (c) cos x, (d) x cos x, (e) $1/(1+x^2)$,
   (f) $\cos(x^2)$, (g) $x^2\cos \frac{1}{x}$, (h) $x^2\sin \frac{1}{x}$.

18. Prove that $\cos(x^2)$ is <u>not</u> uniformly continuous
on R.

19. The real function f is defined and is continuous on $]0,1]$.   It is given that
            sup f = 1 and inf f = -1,

but that neither sup f nor inf f is attained.
Prove, by a contradiction argument or otherwise, that
(i) for every $\delta > 0$, there exists $x \varepsilon]0,\delta[$ such that
$f(x) > \frac{1}{2}$, and (ii) for every $\delta > 0$, there exists
$y \varepsilon]0,\delta[$ such that $f(y) < -\frac{1}{2}$.

   Deduce that f oscillates boundedly as $x \to 0+$.

20.   Show that Q with $d(x,y) = |x-y|$ is a metric
space.   Show that Q is <u>not</u> complete, by exhibiting a
Cauchy sequence in Q which has no limit in Q.

21.   A metric space (S,d) is called <u>bounded</u> if there
exists a constant K such that $d(x,y) \leq K$ for all
$x,y \varepsilon S$.   Prove that a compact metric space is bounded.
[Hint: Use a contradiction argument.]

22.   Let S = C[0,1], the set of all continuous
functions on [0,1].   Define d on S × S by
$$d(f,g) = \sup_{x \varepsilon [0,1]} |f(x)-g(x)|.$$
Prove that d is well-defined (i.e. the supremum does
exist) and that (S,d) is a metric space.

23.   Generalise Ex.3.28 to a function $f: S \to R$,
where S is a general metric space.

24.   The function f is continuous on R.   Also
$f(x) \to 0$ as $x \to \pm\infty$.   Prove that f is uniformly
continuous on R.

25.   Prove that a function $f: R \to R$ which is
continuous and periodic is bounded.

V. THE DERIVATIVE AND SOME APPLICATIONS

---

119. DERIVATIVE

Let f be a real function defined on a neighbourhood of a point a.   Then if the limit

$$\lim_{x \to a} \frac{f(x) - f(a)}{x - a}$$

exists, it is denoted by f'(a) and is called the
derivative of f at a.   The function f is then
differentiable at a.   (Similarly for derivatives on the
left and right, denoted by $f'_L(a)$ and $f'_R(a)$.)

We say f is differentiable on an interval I if
f is differentiable at every interior point of I and
on the left/right at the endpoints if they are in I.
The set of all values f'(a) then defines a function
f' on I, called the derivative of f.

In general terms f is differentiable at a if
a non-vertical tangent can be drawn to the graph of f
at the point (a,f(a)); then f'(a) is the gradient of
the tangent to the curve y = f(x) at this point.

Finding the derivative of f' (if this is possible)
gives f", the second derivative of f.   Further
differentiation (if possible) gives derivatives of
higher order, f''' and so on.

Notice also the $\frac{d}{dx}$ notation for the derivative in
the following example.

EXAMPLE.   Functions f, g are defined on R by

$$f(x) = x^n \quad \text{(for some } n \varepsilon N), \quad g(x) = 5|x|.$$

Prove that $f'(x) = nx^{n-1}$ for all $x \varepsilon R$, but that g'(0)
does not exist.

Solution.   Let $a \varepsilon R$.   Then

$$\frac{f(x) - f(a)}{x - a} = \frac{x^n - a^n}{x - a} = x^{n-1} + x^{n-2}a + \ldots + a^{n-1}$$

$$\to na^{n-1} \quad \text{as} \quad x \to a.$$

So $f'(a) = na^{n-1}$, or in other notation $\frac{d}{dx}(x^n) = nx^{n-1}$.

Also $\frac{g(x) - g(0)}{x - 0} = \frac{5|x|}{x} = \begin{cases} 5 & (x > 0), \\ -5 & (x < 0). \end{cases}$

Since this has no limit as $x \to 0$, g'(0) does not exist.

120. DIFFERENTIABILITY IMPLIES CONTINUITY

RESULT. If f is differentiable at a, then f is continuous at a. (Similarly on the left/right.)

Proof. Suppose that f is differentiable at a. Then

$$f(x) - f(a) = \frac{f(x) - f(a)}{x - a} \cdot (x - a) \to f'(a) \cdot 0 = 0,$$

as $x \to a$. So $f(x) \to f(a)$ as $x \to a$, i.e. f is continuous at a.

N.B. The converse of this result is false: e.g. $5|x|$ in §119 is continuous but not differentiable at 0.

121. INHERITANCE OF DIFFERENTIABILITY UNDER SCALAR MULTIPLICATION, ADDITION, MULTIPLICATION, DIVISION

RESULT. Let f, g be differentiable at a. Then so also are the functions

kf ($k \varepsilon R$), $f \pm g$, fg and f/g, (the quotient with the proviso that it is defined, i.e. $g(a) \neq 0$), with derivatives $kf'(a)$, $f'(a) \pm g'(a)$, $f'(a)g(a) + f(a)g'(a)$, and $(f'(a)g(a) - f(a)g'(a))/(g(a))^2$.

Proof. (e.g. for the product)

$$\frac{f(x)g(x) - f(a)g(a)}{x-a} = \frac{f(x)g(x) - f(a)g(x)}{x-a} + \frac{f(a)g(x) - f(a)g(a)}{x-a}$$

$$= \frac{f(x) - f(a)}{x-a} \cdot g(x) + f(a) \cdot \frac{g(x) - g(a)}{x-a}$$

$$\to f'(a)g(a) + f(a)g'(a) \quad \text{as} \quad x \to a$$

(using the continuity of g at a).

Notice that together with the example of §119 this at once gives derivatives for polynomials and rational functions wherever they are defined.

122. INHERITANCE OF DIFFERENTIABILITY UNDER COMPOSITION OF FUNCTIONS — THE CHAIN RULE

Recall from §83 that $f \circ g$ denotes the composition of the functions f and g and that $(f \circ g)(x) = f(g(x))$. If we write $h = f \circ g$, the following result can be written in d/dx notation as

$$\frac{dh}{dx} = \frac{dh}{dg} \cdot \frac{dg}{dx} .$$

RESULT. Suppose that g is differentiable at c and that f is differentiable at $g(c)$. Then $f_{o}g$ is differentiable at c and

$$(f_{o}g)'(c) = f'(g(c)).g'(c). \qquad \ldots(*)$$

Proof. The proof splits into cases 1,2 and 3 according as the first, the second or neither of the following conditions is satisfied:

$$\exists \text{ neighbourhood } U \text{ of } c \text{ such that}$$
$$\forall x \in U-\{c\}, \quad g(x) \neq g(c). \qquad \ldots(\alpha)$$

$$\exists \text{ neighbourhood } U \text{ of } c \text{ such that}$$
$$\forall x \in U-\{c\}, \quad g(x) = g(c). \qquad \ldots(\beta)$$

In cases 2 and 3 we find that both sides of (*) take the value zero.

Case 1. (The straightforward case) Suppose that condition (α) is satisfied. Then, for all $x \in U-\{c\}$, since $g(x) \neq g(c)$ we can write $g(x) - g(c)$ in the denominator and see that

$$\frac{f(g(x)) - f(g(c))}{x - c} = \frac{f(g(x)) - f(g(c))}{g(x) - g(c)} . \frac{g(x) - g(c)}{x - c}$$
$$\rightarrow f'(g(c)).g'(c) \quad \text{as} \quad x \rightarrow c.$$

[The fact that g is continuous at c (because it is differentiable) is used to see that $g(x) \rightarrow g(c)$ as $x \rightarrow c$.]

So (*) is proved in this case.

Case 2. (The trivial case) Suppose in this case that condition (β) is satisfied. Then g is constant on a neighbourhood of c and consequently $f_{o}g$ is constant on a neighbourhood of c. So $g'(c) = 0$, $(f_{o}g)'(c) = 0$ and (*) is therefore proved, because both sides are zero.

Case 3. (The remaining case) Suppose now that neither (α) nor (β) is satisfied. This means (by the rules for negation of quantifiers of §32) that

$$\forall \text{ neighbourhoods } U \text{ of } c,$$
$$\exists x \in U-\{c\} \text{ such that } g(x) = g(c), \qquad \ldots(\alpha')$$

and $\forall$ neighbourhoods U of c,
$$\exists x \in U-\{c\} \text{ such that } g(x) \neq g(c). \qquad \ldots(\beta')$$

For some fixed neighbourhood V of c, write $A = \{x \in V-\{c\} : g(x) = g(c)\}$, $B = \{x \in V-\{c\} : g(x) \neq g(c)\}$.

From (α') and (β') it follows that we can find sequences tending to c that lie entirely in A and also sequences tending to c that lie entirely in B. Let $\{x_n\}$ be such a sequence that lies entirely in A. Then notice that, since $g'(c)$ is known to exist, its value can be obtained from

$$g'(c) = \lim_{x_n \to c} \frac{g(x_n) - g(c)}{x_n - c} = \lim_{x_n \to c} \frac{0}{x_n - c} = 0. \quad \ldots(1)$$

This means that the right-hand side of (*) is zero. ..(2)

Notice also that

$$\frac{f(g(x_n)) - f(g(c))}{x_n - c} = \frac{0}{x_n - c} \to 0 \quad \text{as} \quad x_n \to c. \quad \ldots(3)$$

Now take a sequence $\{y_n\}$ tending to c but lying entirely in B. It follows that

$$\frac{f(g(y_n)) - f(g(c))}{y_n - c} = \frac{f(g(y_n)) - f(g(c))}{g(y_n) - g(c)} \cdot \frac{g(y_n) - g(c)}{y_n - c}$$

$$\to f'(g(c)).g'(c) \quad \text{as} \quad y_n \to c,$$
$$= 0 \quad \text{(by (1) above)}. \quad \ldots(4)$$

Since $V - \{c\} = A \cup B$, it follows from (3) and (4) that

$$\lim_{x \to c} \frac{f(g(x)) - f(g(c))}{x - c} = 0,$$

so that the left-hand side of (*) exists and is 0. ..(5)

Lines (2) and (5) give the result (*) as required.

N.B. Most practical applications of the above result fall within the scope of Case 1. However, to see how Case 3 can arise, look at the situation where $c = 0$ and g is defined on R by

$$g(0) = 0 \quad \text{and} \quad g(x) = x^2 \sin(1/x) \quad (x \neq 0).$$

[For the graph of g, look at §126.]

123. INHERITANCE OF DIFFERENTIABILITY BY THE INVERSE FUNCTION

RESULT. Suppose that f is a strictly monotonic function on an interval I and suppose that f is differentiable at $c \in I$. Let $f(c) = \gamma$. Then, provided that $f'(c) \neq 0$, f^{-1} is differentiable at γ and its derivative there is given by

$$(f^{-1})'(\gamma) = 1/f'(c).$$

Proof.
$$\frac{f^{-1}(y) - f^{-1}(\gamma)}{y - \gamma} = \frac{f^{-1}(y) - f^{-1}(\gamma)}{f(f^{-1}(y)) - f(f^{-1}(\gamma))}$$

$$= \frac{x - c}{f(x) - f(c)} \text{, on writing } x = f^{-1}(y).$$

Now recall that, since f is continuous at c, so also is f^{-1} at γ, by the result of §89. So, as $y \to \gamma$, $f^{-1}(y) \to f^{-1}(\gamma)$, i.e. $x \to c$. So,

$$\lim_{y \to \gamma} \frac{f^{-1}(y) - f^{-1}(\gamma)}{y - \gamma} = \lim_{x \to c} \frac{x - c}{f(x) - f(c)}$$

$$= \frac{1}{f'(c)} \quad \text{(provided that } f'(c) \neq 0).$$

124. DIFFERENTIATION OF FRACTIONAL POWERS OF X

Differentiation of positive integral powers of x has been covered in §119. Recall that the function f defined by $f(x) = x^n$ ($n \varepsilon N$) has inverse $f^{-1}(y) = y^{1/n}$, for suitable domains. The result of §123 finds $(f^{-1})'(y)$: $(f^{-1})'(y) = 1/f'(x) = 1/(nx^{n-1}) = \frac{1}{n} y^{1/n-1}$ where $y = x^n$ and provided $y \neq 0$.

Working from this basic result for the differentiation of $y^{1/n}$ and using the product and quotient rules for differentiation, we can then prove that, for suitable values of y and for every rational number α, $\frac{d}{dy}(y^\alpha) = \alpha y^{\alpha-1}$.

Derivatives at 0 are not covered by these arguments and direct calculation of derivatives reveals various possibilities: e.g. $x^{4/3}$ is differentiable at 0, $x^{3/2}$ is differentiable on the right at 0 (only defined for $x > 0$), and $x^{1/3}$ is not differentiable at 0 since the tangent to the graph is vertical.

125. EXAMPLE. (on the use of §123) Look at the case

of the differentiable, strictly increasing function sin with domain $[-\frac{\pi}{2},\frac{\pi}{2}]$ and range $[-1,1]$.

Then the inverse function $\sin^{-1}$ with domain $[-1,1]$ and range $[-\frac{\pi}{2},\frac{\pi}{2}]$ has derivative (from §123) given by

$$\frac{d}{dy}(\sin^{-1}y) = \frac{1}{\frac{d}{dx}(\sin x)} \quad , \quad \text{where} \quad y = \sin x,$$

$$= \frac{1}{\cos x} = \frac{1}{\sqrt{(1-\sin^2 x)}} = \frac{1}{\sqrt{(1-y^2)}}$$

(provided that $\cos x \neq 0$, i.e. $y \neq \pm 1$). See also §219.

126. FUNCTIONS WITH DISCONTINUOUS DERIVATIVE

If a function is differentiable on an interval, it might seem reasonable that its derivative should be continuous on the interval. However this is <u>not</u> the case as the following example shows.
Look at f defined on R by

$$f(x) = x^2 \sin \frac{1}{x} \quad (x \neq 0),$$

$$f(0) = 0.$$

The graph of this function lies between the parabolas $y = \pm x^2$ as shown.

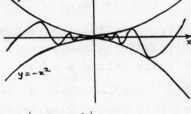

Here, $f'(0) = \lim_{x \to 0} \dfrac{x^2 \sin \frac{1}{x} - 0}{x - 0}$

$= \lim_{x \to 0} x \sin \frac{1}{x} = 0$, (since $\left| x \sin \frac{1}{x} \right| \leq |x|$).

Suppose now that f' is continuous at $x = 0$.
Then there exists $\delta > 0$ such that

$$|f'(x)| < \tfrac{1}{2} \quad \forall x \in \,]-\delta, \delta[. \qquad \ldots (*)$$

Now choose $k \varepsilon N$ with $0 < 1/(2k\pi) < \delta$. Then, if $x \neq 0$,

$$f'(x) = 2x \sin \frac{1}{x} - \cos \frac{1}{x} \, , \text{ by the product rule etc.}$$

So $f'(1/(2k\pi)) = -\cos 2k\pi = -1$. This contradicts $(*)$. So the above assumption is wrong. So f' is <u>not</u> continuous at $x = 0$.

(Notice also that $f'(1/(2k+1)\pi) = 1$ for all $k \varepsilon N$.)

<u>EXAMPLE</u>. g is defined on R by

$$g(x) = x \sin \frac{1}{x} \quad (x \neq 0), \quad g(0) = 0.$$

Draw a rough graph of g and show that g is continuous but is not differentiable at $x = 0$.

127. MAXIMA, MINIMA, POINTS OF INFLEXION

The condition $f'(a) = 0$ is necessary for the point $(a, f(a))$ to be a turning point (i.e. a maximum or a minimum) on the graph of a function differentiable at $x = a$. However this condition (i.e. $f'(a) = 0$) is not sufficient for a turning point (e.g. look at x^3 at $x = 0$). The condition $f''(a) = 0$ is necessary for $(a, f(a))$ to be a point of inflexion on the graph of a function twice differentiable at $x = a$, but it is not sufficient (e.g. x^4 at $x = 0$). At a point of inflexion the sign of f'' changes. See books on elementary calculus.

128. ROLLE'S THEOREM

RESULT. Let the real function f be continuous on $[a,b]$, differentiable on $]a,b[$, and let $f(a) = f(b)$. Then there exists a point $c \in]a,b[$ with $f'(c) = 0$.

Proof. Suppose that f is not constant. (The case of a constant function is trivial.) Then f is bounded and attains its bounds by §107, and at least one of sup f and inf f is attained on the open interval $]a,b[$. Suppose without loss of generality that sup f is attained at $c \in]a,b[$. So $f(x) \leq f(c)$ $\forall x \in [a,b]$. So

$$f'_L(c) = \lim_{x \to c^-} \frac{f(x) - f(c)}{x - c} \geq 0$$

and

$$f'_R(c) = \lim_{x \to c^+} \frac{f(x) - f(c)}{x - c} \leq 0.$$

But $f'(c) = f'_L(c) = f'_R(c)$. So $f'(c) = 0$.

Notice that this result can be useful in locating zeroes of functions. For example, let $p(x) = (x-2)^3(x-6)(x-9)$. Then $p(2) = p(6) = p(9) = 0$. So deduce from Rolle that p' has a zero in $]2,6[$ and another in $]6,9[$.

129. THE MEAN-VALUE THEOREM

This useful result is easy to apply.

RESULT. Let f be a real function, continuous on
$[a,b]$ and differentiable on $]a,b[$. Then there exists
a point $\xi \epsilon]a,b[$ such that
$$f(b)-f(a) = (b-a)f'(\xi).$$

Proof. Let $G(x) = f(x)-\alpha x$ where α is a constant
chosen to make $G(a) = G(b)$, i.e. $\alpha = \dfrac{f(b)-f(a)}{b-a}$. Then
Rolle's theorem applies to G on $[a,b]$ and we deduce
the existence of a point $\xi \epsilon]a,b[$ such that $G'(\xi) = 0$,
i.e. $(f'(\xi)-\alpha) = 0$. So $(f(b)-f(a)) = (b-a)f'(\xi)$ as
required.

Notice: (i) The mean-value theorem estimates the
height difference between the points $(a,f(a))$ and
$(b,f(b))$ on the graph of the function f. The result
effectively says that

| height difference i.e. $(f(b)-f(a))$ | = | horizontal distance i.e. $(b-a)$ | × | average rate of climb i.e. $f'(\xi)$ | . |

(ii) The estimate of $f(b)-f(a)$ is made assuming only
the existence of f'. The more refined estimates encom-
passed by Taylor's theorem (§172) are available in cases
where the existence of more derivatives is known. However, even in such cases the mean-value theorem often
leads to worthwhile conclusions. See for example §131,
§134 and §179.

130. EXAMPLE ON MONOTONIC FUNCTIONS

EXAMPLE. Let f be a real function defined on an open
interval I. Suppose that $f'(x) > 0$ on I. Then f
is strictly increasing on I.

Solution. Take $a,b\epsilon I$ with $a < b$. The mean-value
theorem gives
$$f(b)-f(a) = (b-a)f'(\xi) > 0.$$
So $f(b) > f(a)$ and so f is strictly increasing on I.

(Similarly f is constant on I if $f'(x) = 0$ on I
and f is strictly decreasing on I if $f'(x) < 0$ on I.
Also, if $f'(x) > 0$ on $]a,b[$ and f is continuous
on $[a,b]$ then f is strictly increasing on $[a,b]$.)

D

131.　**EXAMPLE.**　Prove that $\sin x < x < \tan x$ on $]0,\frac{\pi}{2}[$.

Solution.　Define f and g on $[0,\frac{\pi}{2}[$ by $f(x) = x - \sin x$ and $g(x) = \tan x - x$. Then,

$$f'(x) = 1 - \cos x > 0 \quad \text{on} \quad]0,\frac{\pi}{2}[,$$

$$g'(x) = \sec^2 x - 1 = \frac{1 - \cos^2 x}{\cos^2 x} > 0 \quad \text{on} \quad]0,\frac{\pi}{2}[.$$

So, by §130, f and g are strictly increasing on $[0,\frac{\pi}{2}[$. But $f(0) = g(0) = 0$. So $f(x) > 0$ and $g(x) > 0$ on $]0,\frac{\pi}{2}[$, i.e. $x > \sin x$ and $\tan x > x$ as required.

EXAMPLE.　Prove that

(i) $\cos x > 1 - \frac{1}{2}x^2$ on $]0,\frac{\pi}{2}[$,

(ii) $\log(1+x) > x - \frac{1}{2}x^2$ on $]0,\infty[$.

132.　**EXAMPLE.**　Prove that each of the equations
$x^5 + 2x - 2 = 0$ and $x^3 + x^2 + x - 5 = 0$ has only one real root.

Solution.　Using the intermediate value theorem as in Ex.3.11 we are guaranteed at least one real root for eac equation. Let f,g be defined on R by $f(x) = x^5 + 2x -$ $g(x) = x^3 + x^2 + x - 5$. Notice that $f'(x) = 5x^4 + 2 > 0$ on Also $g'(x) = 3x^2 + 2x + 1 > 0$ on R, because in the notati of §29 $b^2 - 4ac = -8$. So f and g are strictly increasing on R and so can take each real value at mos once. So each equation has exactly one real root.

133.　**FUNCTIONS WHICH MAINTAIN A RATE OF CLIMB**

EXAMPLE.　(i) Let $f: R \to R$ be differentiable on R and suppose that $f'(x) \geq K > 0$ for all $x > N$. Then $f(x) \to \infty$ as $x \to \infty$. (Similarly, if $f'(x) < k < 0$ for all $x > N$, $f(x) \to -\infty$ as $x \to \infty$.)

Solution.   By the mean-value theorem, for $x > N$,
$$f(x)-f(N) = (x-N)f'(\xi) \quad \text{where} \quad \xi\epsilon\,]N,x[.$$
$$\text{So} \quad f(x) \geq f(N) + (x-N)K \to \infty \quad \text{as} \quad x \to \infty.$$

EXAMPLE.   (ii) Let $f: R \to R$ be differentiable on R, and suppose that $f(x) \to A$ and $f'(x) \to B$ as $x \to \infty$. Prove that $B = 0$.

Solution.   Suppose that $B \neq 0$.   Then either $B > 0$ or $B < 0$.   Suppose without loss of generality that $B > 0$. Then since $f'(x) \to B$ as $x \to \infty$,
$$\exists X \quad \text{such that} \quad |f'(x)-B| < \tfrac{1}{2}B \quad \forall x \geq X.$$
$$\text{So} \quad f'(x) > \tfrac{1}{2}B \quad \forall x \geq X.$$
It follows from (i) that $f(x) \to \infty$ as $x \to \infty$, contradicting the fact that $f(x) \to A$ as $x \to \infty$.   So our original assumption was wrong and $B = 0$.

134.   ESTIMATION OF JUMP IN VALUE

The jump in the value of a function on an interval can often be estimated by the mean-value theorem, as the following examples illustrate:

EXAMPLE.   (i) Estimate $\sqrt{(n+5)} - \sqrt{n}$ where $n\epsilon N$, (i.e. estimate the jump of the function $\sqrt{x}$ on $[n,n+5]$).

Solution.   Applying the mean-value theorem to the function f defined by $f(x) = \sqrt{x}$ on $[n,n+5]$ gives
$$\sqrt{(n+5)} - \sqrt{n} = 5.\frac{1}{2\sqrt{\xi}} \quad \text{where} \quad n < \xi < (n+5).$$
So, $\dfrac{5}{2\sqrt{(n+5)}} < \sqrt{(n+5)} - \sqrt{n} < \dfrac{5}{2\sqrt{n}}.$

For example, if we take $n = 100$ and notice that $\sqrt{105} < 11$, we can conclude that $\sqrt{105}$ lies between $10\frac{5}{22}$ and $10\frac{1}{4}$.

EXAMPLE.   (ii) Prove that, for every $n\epsilon N$,
$$\frac{1}{n+1} < \log(n+1) - \log n < \frac{1}{n}.$$

Solution. Applying the mean-value theorem to $\log x$ on $[n, n+1]$ gives

$$\log(n+1) - \log n = 1 \cdot \frac{1}{\xi} \quad \text{where} \quad n < \xi < n+1.$$

So, $\frac{1}{n+1} < \log(n+1) - \log n < \frac{1}{n}$ as required.

Many other cases of such estimation occur later, e.g. in §136, §179(ii), Ex.8.1, Ex.8.3.

135. CONTRACTION MAPPINGS

Let I be a subinterval of R. Then a function $f: I \to I$ is a <u>contraction mapping</u> if

$$|f(x) - f(y)| < |x - y| \quad \text{for all} \quad x, y \in I.$$

(In other words the action of f on a pair of points decreases the distance between them.) There is an analogous definition for a general metric space.

<u>EXAMPLE.</u> (i) Suppose f is a differentiable function on an open interval I and that $|f'(x)| < 1$ for all $x \in I$. Prove that there is at most one root of the equation $f(x) = x$ in I.

<u>Solution.</u> Suppose that a, b (with $a < b$) are two roots of the equation. Then, by the mean-value theorem,

$$(b-a) = f(b) - f(a) = f'(\xi)(b-a) < (b-a),$$

because $f'(\xi) < 1$. This is a contradiction. So the original assumption of two distinct roots is wrong.

<u>EXAMPLE.</u> (ii) Show that the equation $x^3 + 8x - 10 = 0$ has exactly one real root α and that $1 < \alpha < 2$. A sequence $\{a_n\}$ is defined by taking a_1 arbitrarily in $[1,2]$ and $a_{n+1} = f(a_n)$ for $n \geq 1$, where $f(x) = \frac{10}{x^2+8}$. Prove that $|f'(x)| < \frac{15}{16}$ for all $x \in]0,3[$ and use the mean-value theorem to show that $|a_{n+1} - \alpha| < \frac{15}{16}|a_n - \alpha|$. Deduce that $a_n \to \alpha$ as $n \to \infty$. Use a calculator to demonstrate that $\alpha = 1.089$ to three decimal places.

Solution. Define g on R by $g(x) = x^3 + 8x - 10$. Notice that g is continuous on R, that $g(1) = -1$ and that $g(2) = 14$. So the intermediate value theorem guarantees <u>at least one</u> root of the equation $g(x) = 0$ in the interval $]1,2[$. Furthermore $g'(x) = 3x^2 + 8 > 0$ for all $x \varepsilon R$. So g is strictly increasing on R and there must be <u>exactly one</u> root of $g(x) = 0$ in $]1,2[$. Denote the root by α.

On differentiation notice that

$$|f'(x)| = \frac{20|x|}{(x^2+8)^2} < \frac{60}{64} = \frac{15}{16} \quad \text{for all} \quad x \varepsilon]0,3[.$$

Then, noticing that $f(\alpha) = \alpha$ and using the mean-value theorem, see that

$$|a_{n+1} - \alpha| = |f(a_n) - f(\alpha)| = |f'(\xi)||a_n - \alpha|$$
$$< \frac{15}{16}|a_n - \alpha| \quad \text{if} \quad a_n \varepsilon [0,3]. \qquad \ldots (*)$$

Suppose now that the distance of a_n from α is less than 1. Certainly then $a_n \varepsilon [0,3]$ and $(*)$ guarantees that the distance of a_{n+1} from α is less than 15/16. So then a_{n+1} is closer to α than a_n and so also $a_{n+1} \varepsilon [0,3]$. Clearly we can use $(*)$ again to repeat the process and show that a_{n+2} is closer to α than a_{n+1} and so on.

If we start with $n = 1$ and $a_1 \varepsilon [1,2]$ as suggested in the question, we can use the above argument repeatedly (in a formal induction argument if desired) to see that $|a_{n+1} - \alpha| < \left(\frac{15}{16}\right)^n |a_1 - \alpha|$. It follows from this that $a_n \to \alpha$ as $n \to \infty$, since $(15/16)^n \to 0$ as $n \to \infty$.

(If $a_1 = 1.5$, sucessive terms are $1.500, 0.976, 1.117, 1.081, 1.091, 1.088, 1.089, 1.089, \ldots$. Notice the oscillatory approach to the limit.)

136. EXAMPLE ON UNIFORM CONTINUITY

<u>EXAMPLE</u>. Let f be a real function differentiable on R and such that there exists a constant $K > 0$ such that $|f'(x)| \le K$ for all $x \varepsilon R$. Prove that f is uniformly continuous on R.

Solution. Let $\varepsilon > 0$. Then, for $x, y \varepsilon R$, the mean-value theorem gives $|f(x)-f(y)| = |(x-y)f'(\xi)| \leq K|x-y| < \varepsilon$ for all x, y with $|x-y| < \frac{\varepsilon}{K}$. So f is uniformly continuous on R.

The above example gives a useful criterion for checking uniform continuity on an interval which is not closed; in such cases the criterion of §111 is inapplicable. As examples, notice that the above example guarantees the uniform continuity of x, sin x and $1/(1+x^2)$ on R.

137. EXAMPLE. Show that the sequence $\{f(n)\}$ defined by $f(n) = \frac{(\log n)^2}{n}$ is eventually decreasing.

Solution. (The method of §65 is not helpful here.) Think of f as a real function defined on $]0, \infty[$. Then

$$f'(x) = \frac{2 \log x - (\log x)^2}{x^2} < 0 \quad \text{for} \quad x \geq N, \text{ say.}$$

(The numerator is dominated by the $-(\log x)^2$ term as $x \to \infty$.) So since f' is negative for $x \geq N$ conclude that f decreases for $x \geq N$. Hence the result. (The sequence actually tends to zero by the hierarchy result of §209.)

138. THE INTERMEDIATE VALUE PROPERTY FOR DERIVATIVES: THE RESULT OF DARBOUX

Derivatives need not be continuous as the example of §126 shows. Nevertheless, they do possess the intermediate value property possessed by continuous functions. See Ex.5.26.

EXAMPLES 5
(All numbers in these examples are real.)

1. Find the derivatives of x^2, $1/x$ and $1/(x+2)$ from first principles.

2. Let $f(x) = 1 + \frac{1}{2}x - \sqrt{(1+x)}$ for all $x \geq 0$. Show that $f(0) = 0$ and by considering f' show that f is strictly increasing on $[0, \infty[$. Deduce that $1 + \frac{1}{2}x > \sqrt{(1+x)}$ for all $x > 0$.

3. Prove that each of the equations $x^3 + 5x - 15 = 0$ and $x^5 - x^3 + x - 12 = 0$ has only one real root. Show that both roots lie in $]1,2[$.

4. Prove that $\sin x > x - \frac{x^3}{6}$ for all $x \varepsilon]0, \frac{\pi}{2}[$. (You may need the result of Example (i) in §131.)

5. Let f be twice differentiable on R and suppose that $f(x) \to \infty$ as $x \to \infty$ and $f(x) \to -\infty$ as $x \to -\infty$. Suppose also that there exist two points a,b with $a < b$ such that $f(a) > 0$ and $f(b) < 0$. Use Rolle's result to show that there exists $c \varepsilon R$ such that $f''(c) = 0$. Give an explicit example of such a function f.

6. Let $a,b \varepsilon R$ with $a < b$. Apply Rolle's result to the function f defined by $f(x) = e^{-x}(x-a)(x-b)$ to show that the equation

$$(x-a)(x-b) = (x-a) + (x-b)$$

has a root between a and b.

7. Let f be a periodic infinitely differentiable function with period p. (Examples are $\sin x$ and $\cos x$.) Use Rolle's result to prove that for every positive integer n there exists a point $x_n \varepsilon [0,p[$ such that $f^{(n)}(x_n) = 0$.

8. Give an example to show that a function different-iable on $]0,\infty[$ and for which $f'(x) > 0$ for all $x > N$, need not tend to infinity as $x \to \infty$. (Contrast this with the effect of the stronger condition in §133(i).)

9. The function f is differentiable on R and $f'(x) \to A$ as $x \to \infty$, where $A \neq 0$. Prove that $f(x) \to \infty$ or $f(x) \to -\infty$ as $x \to \infty$.

10. Suppose that $f''(x) > 0$ for all $x \varepsilon R$.   Prove
that $f'(x)$ either tends to a limit or to infinity as
$x \to \infty$.   Give examples to show that both possibilities
can occur.   (The sign of f'' can be related to the
<u>concavity</u> of the graph of f.   If f'' is positive on
an interval, the graph is concave up.)

11. A positive function f is differentiable on R
and $f(x) \to 0$ as $x \to \pm\infty$.   It then follows from Ex.4.11
that f attains its supremum at a point c.   Prove
that $f'(c) = 0$.

12. The function g is twice differentiable on R.
Also g is bounded on R and $g''(x) \to B$ as $x \to \infty$.
Prove that $B = 0$.

13. (Compare §135.)   Show that the equation
$x^4+3x-1 = 0$ has one root $\alpha \varepsilon]0,\frac{1}{2}[$.   A sequence $\{a_n\}$
is defined by choosing a_1 arbitrarily in $]0,\frac{1}{2}[$ and
$a_{n+1} = f(a_n)$ for all $n \varepsilon N$, where $f(x) = (x^3+3)^{-1}$.
Prove that $|f'(x)| < \frac{3}{4}$ for all $x \varepsilon]-1,1[$, and use the
mean-value result to prove that $a_n \to \alpha$ as $n \to \infty$.
Use a calculator to show that $\alpha = 0\cdot3294$ (4 places).

14. Show that there is a root α of the equation
$x^3+12x-15 = 0$ in the interval $]1,2[$.   Devise and
justify an algorithm similar to that in Ex.13 for finding
α and use a calculator to show that $\alpha = 1\cdot130$ (3 places).

15. Show that the equation $x^3+x-60 = 0$ has exactly
one real root α and that $\alpha \varepsilon]3,4[$.   A sequence is
defined by taking a_1 arbitrarily in [3,4] and
$a_{n+1} = f(a_n)$ for $n \varepsilon N$, where $f(x) = \dfrac{60}{x^2+1}$.   Show that
$|f'(x)| > 1$ for all $x \varepsilon]3,4[$ and deduce using the mean-
value theorem that a_2 lies further from α than a_1.
Deduce that $a_n \not\to \alpha$ as $n \to \infty$.   Use a calculator to
examine the behaviour of $\{a_n\}$ as $n \to \infty$.

16. (<u>To find</u> α in Ex.15.)   A sequence $\{a_n\}$ is
defined by taking a_1 arbitrarily in [3,4] and
$a_{n+1} = g(a_n)$ for $n \varepsilon N$, where $g(x) = \dfrac{30(x+2)}{x^2+31}$.   Find a

constant k with $0 < k < 1$, such that $|g'(x)| < k$ for all $x \in [2,5]$ and deduce that $a_n \to \alpha$, the real root of the equation $x^3 + x - 60 = 0$. Use a calculator to show $\alpha = 3 \cdot 8297$ (4 places).

17. (<u>Newton's method for finding a root of an equation</u>) Suppose we are to find a root α of the equation $h(x) = 0$ within an interval I, on which h' exists and is non-zero. Define a sequence by choosing a_1 arbitrarily in $[\alpha-s, \alpha+s] \subseteq I$, and then $a_{n+1} = f(a_n)$ for $n \in N$, where $f(x) = x - \dfrac{h(x)}{h'(x)}$. Prove that $f'(x) = \dfrac{h(x)h''(x)}{(h'(x))^2}$ and use the method of Example (ii) of §135 to show that if there exists a constant k with $0 < k < 1$ such that

$$|h(x)h''(x)| < k(h'(x))^2 \quad \text{for all} \quad x \in [\alpha-s, \alpha+s]$$

then $a_n \to \alpha$ as $n \to \infty$.

Use a calculator and the above method to locate the root of the equation $x^3 + x - 60 = 0$ in the interval $[3,4]$. Compare the result with that of Ex.16.

18. The function g is differentiable on R and $g'(x) \to \infty$ as $x \to \infty$. Use the mean-value theorem to prove that g is <u>not</u> uniformly continuous on R. Hence show that x^2 and x^3 are not uniformly continuous on R.

19. Use Ex.4.24 to show that the function f defined on R by $f(x) = \dfrac{\sin(x^4)}{x^2+1}$ is uniformly continuous on R. Show however that f' is unbounded on R. (So a uniformly continuous differentiable function need not have bounded derivative on the interval in question.)

20. The function f is differentiable on R and there exists a constant K such that, for all $x, y \in R$, $|f(x) - f(y)| < K|x-y|$. Show that f' is bounded on R. Deduce that f is uniformly continuous on R.

Deduce from this and Ex.19 that if g is uniformly continuous on R, it does <u>not</u> follow that there exists a number K such that for all $x, y \in R$, $|g(x) - g(y)| < K|x-y|$.

21. Give an example to show that if f is differentiable on]0,∞[and oscillates boundedly as x → ∞, then f' can still tend to zero as x → ∞. (Use trigonometric functions.)

22. Give an example involving trigonometric functions to show that if f is differentiable on]0,∞[and f(x) → A as x → ∞, it is possible that f' oscillates boundedly as x → ∞.   (Compare §133(ii) where slightly stronger conditions ensure that f'(x) → 0 as x → ∞.)

23. The function f is differentiable on R and f'(x) ≥ f(x) > 0 for all x∈R.   Prove that f(x) → ∞ as x → ∞.

24. The function g is twice differentiable on R. Also g is bounded on R and g''(x) > 0 for all x∈R. Prove that g' is increasing and bounded above on R. Deduce that g'(x) → 0 as x → ∞.
   Prove further that g(x) tends to a limit as x → ∞.

25. Show that if f is differentiable on [a,∞[and f'(a) < 0, then there exists a point d > a such that f(d) < f(a).

26. (<u>Darboux's result: see §138.</u>) Let f be differentiable on [a,b] and let f'(a) < 0, f'(b) > 0. Show that   inf f is attained at a point c∈]a,b[and
          [a,b]
that f'(c) = 0.
   Deduce that if g is differentiable on [a,b], then g' takes every value between g'(a) and g'(b) between a and b.

VI. SERIES

---

139. A <u>series</u> is an array of real or complex numbers written as

(i) $a_1 + a_2 + a_3 + \ldots$, or as (ii) $\sum a_n$.

For example, in the case when $a_n = 1/n$ for all $n \in N$, the corresponding series appears as

(i) $1 + \frac{1}{2} + \frac{1}{3} + \ldots$, or as (ii) $\sum \frac{1}{n}$.

The number a_n is called the <u>nth term</u> of the series.

Corresponding to every such series there is a sequence $\{s_n\}$ constructed by letting

$$s_n = a_1 + a_2 + a_3 + \ldots + a_n .$$

The series $\sum a_n$ is then called <u>convergent</u> or <u>divergent</u> according as the sequence $\{s_n\}$ is convergent or divergent. (Recall from §49 that the statement that the sequence $\{s_n\}$ is convergent means that it tends to a <u>finite</u> limit as $n \to \infty$. Sequences which fail to meet this requirement are divergent.)

If $s_n \to s$ as $n \to \infty$, then s is called the <u>sum</u> of the series $\sum a_n$, and we write

$$s = \sum_{n=1}^{\infty} a_n .$$

Also, the number s_n is called the <u>nth partial sum</u> of the series. As examples take

$$1 + \frac{1}{2} + \frac{1}{4} + \frac{1}{8} + \ldots \quad \text{(convergent)},$$

$$1 + 1 + 1 + 1 + \ldots \quad \text{(divergent)},$$

$$1 - 1 + 1 - 1 + \ldots \quad \text{(divergent)}.$$

140. For any given series, the primary consideration is usually whether it converges or diverges. If it converges, a secondary consideration may be the value of its sum. Various tests which may settle whether a given series is convergent or divergent are given in §149-§156.

However first notice the geometric series (§144) and the series $\sum 1/n^k$ (§145) as two important points of reference in this whole area. Notice also the elementary but useful observations about dealing with series made in §141-§143.

141. Notice: (i) Changing a finite number of terms in a series does not alter its convergence/divergence.

For suppose that all the changes are made before the Nth term. Then let the original and altered partial sums be s_n and t_n. Then there exists a constant k such that $t_n = s_n + k$ for all $n > N$. So, $\{t_n\}$ tends to a limit if and only if $\{s_n\}$ does.

(ii) Insertion or deletion of a finite number of terms does not alter whether a series converges or diverges.
To see this, argue as in (i).

(iii) If $\sum a_n$ and $\sum b_n$ converge to s and t, then $\sum (ka_n + lb_n)$ converges to $ks + lt$, for $k, l \in C$.
To see this, look at partial sums and apply heredity results for limits.

142. In contrast to (i) and (ii) of §141, notice that making an <u>infinite</u> number of changes in a series may alter its convergence or divergence. Such changes include bracketing, changing, inserting, rearranging or deleting terms in the series. For example,

$$1 - 1 + 1 - 1 + 1 - 1 + \ldots \quad \text{diverges}$$

but on bracketing,

$$(1 - 1) + (1 - 1) + (1 - 1) + \ldots \quad \text{converges.}$$

Again, $1 + \frac{1}{2} + \frac{1}{3} + \frac{1}{4} + \frac{1}{5} + \ldots$ diverges

but on deleting all terms except those with denominator a power of 2,

$$1 + \frac{1}{2} + \frac{1}{4} + \frac{1}{8} + \frac{1}{16} + \ldots \quad \text{converges.}$$

Such changes should therefore be treated with caution.
Notice however that the value of a partial sum (i.e. finitely many terms) is unaffected by bracketing or rearranging.

143. AN ELEMENTARY CHECK

Suppose you are asked whether a given series $\sum a_n$ converges or diverges. If the following simple check can be made easily it is often a good idea to make it at the outset:

Check whether $a_n \to 0$ as $n \to \infty$. If $a_n \nrightarrow 0$, the series diverges. If however $a_n \to 0$, the series may converge or diverge and the check fails.

As justification of the check, suppose that $a_n \not\to 0$ but yet $\sum a_n$ converges so that its partial sum $s_n \to s$ as $n \to \infty$. Then

$$|a_n| = |s_n - s_{n-1}| \leq |s_n - s| + |s - s_{n-1}| \to 0 \quad \text{as } n \to \infty.$$

This means that $a_n \to 0$ as $n \to \infty$. This is a contradiction. So we conclude that if $a_n \not\to 0$ then $\sum a_n$ diverges.

In consequence of this notice that the following series are all divergent:

$$\sum (-1)^n, \quad \sum \cos \frac{1}{n}, \quad \sum \frac{n}{n+1}, \quad \sum n^{1/n}.$$

Be warned however that this check is of very limited use because the majority of series (interesting from our point of view) have the property that $a_n \to 0$ and for these the check fails to settle the question of convergence or divergence. For example, both $\sum 1/n$ and $\sum 1/n^2$ have the property that $a_n \to 0$ as $n \to \infty$, but the former diverges while the latter converges.

SOME STANDARD SERIES

144. THE REAL GEOMETRIC SERIES

This is the series $\sum b^{n-1}$, where $b \varepsilon R$, i.e. the series $1 + b + b^2 + b^3 + \dots$. Here,

$$s_n = 1 + b + b^2 + \dots + b^{n-1} = \frac{1 - b^n}{1 - b} \qquad (b \neq 1)$$

$$= \frac{1}{1 - b} - \frac{b^n}{1 - b} \to \frac{1}{1 - b} \quad \text{when } |b| < 1.$$

If $|b| \geq 1$, then $b^n \not\to 0$ as $n \to \infty$ and the check of §143 shows that the series then diverges.

So the series converges to the sum $1/(1 - b)$ when $|b| < 1$ and diverges when $|b| \geq 1$.

As a minor generalisation notice that $a + ab + ab^2 + \dots$ converges to $a/(1 - b)$ when $|b| < 1$.

145. THE SERIES $\sum 1/n^k$

Here explicit expressions for the partial sums are not easy to find, though certain estimates are found in §179. The action splits into two cases - convergence when $k > 1$ (and the terms are relatively small), divergence when $k \leq 1$ (and the terms are bigger).

Case 1. ($k \leq 1$)

Here
$$s_{2n} - s_n = \frac{1}{(n+1)^k} + \frac{1}{(n+2)^k} + \ldots + \frac{1}{(2n)^k}$$

$$\geq \frac{1}{n+1} + \frac{1}{n+2} + \ldots + \frac{1}{2n}$$

$$\geq \frac{1}{2n} + \frac{1}{2n} + \ldots + \frac{1}{2n} \qquad (n \text{ terms})$$

$$= \frac{1}{2} \qquad \qquad \ldots (*)$$

Suppose now that the series converges to L, say. Then $s_{2n} \to L$ and $s_n \to L$ as $n \to \infty$, so that $(s_{2n} - s_n) \to 0$ as $n \to \infty$. This contradicts $(*)$. So the series diverges for $k \leq 1$. (It is easy to see from §57 that $s_n \to \infty$ as $n \to \infty$.)

Case 2.   ($k > 1$)

Let p be the unique integer such that $2^p \leq n < 2^{p+1}$.

Then
$$s_n \leq 1 + \left(\frac{1}{2^k} + \frac{1}{3^k}\right) + \left(\frac{1}{4^k} + \frac{1}{5^k} + \frac{1}{6^k} + \frac{1}{7^k}\right) + \ldots$$

$$+ \left(\frac{1}{(2^p)^k} + \ldots + \frac{1}{(2^{p+1}-1)^k}\right)$$

$$\leq 1 + \frac{2}{2^k} + \frac{4}{4^k} + \ldots + \frac{2^p}{(2^p)^k}$$

$$= 1 + \frac{1}{2^{k-1}} + \frac{1}{(2^{k-1})^2} + \ldots + \frac{1}{(2^{k-1})^p} \quad .$$

Here the RHS is a partial sum of a convergent geometric series and so the RHS is bounded above. So $\{s_n\}$ is bounded above and increasing and so tends to a limit as $n \to \infty$, i.e. the original series converges for $k > 1$.

146.   THE HARMONIC SERIES: $1 + \frac{1}{2} + \frac{1}{3} + \frac{1}{4} + \ldots$

This divergent series is a particular case of §145 and a very important point of reference in the whole area.   Estimates of its partial sums are made in §180.

147. SERIES WHICH COLLAPSE

The following example illustrates this idea.

__EXAMPLE.__ Show that $\sum\limits_{n=1}^{\infty} \dfrac{1}{4n^2-1} = \dfrac{1}{2}$.

__Solution.__ Using partial fractions notice that

$$\frac{1}{4n^2-1} = \frac{1}{2}\left(\frac{1}{2n-1} - \frac{1}{2n+1}\right) .$$

So, $s_n = \dfrac{1}{2}\left(1 - \dfrac{1}{3} + \dfrac{1}{3} - \dfrac{1}{5} + \dfrac{1}{5} - \dfrac{1}{7} + \ldots + \dfrac{1}{2n-1} - \dfrac{1}{2n+1}\right)$

$\qquad = \dfrac{1}{2}\left(1 - \dfrac{1}{2n+1}\right) \to \dfrac{1}{2}$ as $n\to\infty$.

There are however only a few types of series to which this method will apply.

__EXAMPLE.__ Show that $\sum\limits_{n=1}^{\infty} \dfrac{1}{n(n+2)} = \dfrac{3}{4}$.

SERIES WITH NON-NEGATIVE TERMS

148.
There are really four things that can happen as $n \to \infty$ to the partial sums s_n of a series with __real__ terms, namely, (a) $s_n \to L$ (a finite limit), (b) $s_n \to \infty$, (c) $s_n \to -\infty$, (d) $\{s_n\}$ oscillates boundedly or unboundedly. In case (a) the series is convergent, in cases (b),(c) and (d) it is divergent.

For a series with __non-negative__ terms however, only cases (a) and (b) can occur. This is because $\{s_n\}$ is __increasing__ for such series and the result of §57 then leaves only these two possibilities open.

In the light of this, §149-§152 give some basic methods for settling the convergence/divergence question for series with __non-negative__ terms.

149. COMPARISON

__RESULT.__ Suppose that $0 \leq a_n \leq b_n$ for all $n > X$.

Then (i) $\sum a_n$ converges if $\sum b_n$ converges, and

$\qquad$ (ii) $\sum b_n$ diverges if $\sum a_n$ diverges.

<u>Proof</u>. Alter all terms a_n and b_n with $n \leq X$ to have the value zero. This will (by §141) not alter whether the series converge or diverge but it will ensure that $0 \leq a_n \leq b_n$ for all $n \varepsilon N$. Then if $s_n = a_1 + a_2 + \ldots + a_n$ and $t_n = b_1 + b_2 + \ldots + b_n$, it is clear that $s_n \leq t_n$ for all $n \geq 1$. Remember also that s_n and t_n increase with n. So, if $t_n \to t$ it follows that s_n is increasing and bounded above, and so tends to a limit by §57, which proves (i). On the other hand if $s_n \to \infty$ it follows that $t_n \to \infty$ by §54, which proves (ii).

The comparison technique given by this result is useful but it is often eclipsed by the <u>comparison test</u> of §150.

150. THE COMPARISON TEST

Here a given series $\sum a_n$ is tested against a series of known behaviour like the series of §145.

<u>RESULT</u>. (<u>Comparison test</u>) Suppose that a_n, b_n ($n\varepsilon N$) are positive and that $a_n/b_n \to L$, where $L \neq 0$ as $n \to \infty$. Then $\sum a_n$ and $\sum b_n$ have the same behaviour, i.e. both converge or both diverge.

<u>Proof</u>. Since the ratio tends to L it follows that

$$\tfrac{1}{2}L < a_n/b_n < \tfrac{3}{2}L \quad \text{for all } n > X.$$

$$\text{So,} \quad \tfrac{1}{2}Lb_n < a_n < \tfrac{3}{2}Lb_n \quad \text{for all } n > X.$$

Now apply the result of §149 to the LHS if $\sum b_n$ diverges and to the RHS if $\sum b_n$ converges.

<u>EXAMPLE</u>. Settle whether the following converge :

$$\text{(a) } \sum \frac{n^2+1}{n^5+n+1} \ , \quad \text{(b) } \sum \frac{n}{3n^2-1} \ .$$

<u>Solution</u>. (a) Let $a_n = (n^2+1)/(n^5+n+1)$ and $b_n = 1/n^3$.

Then $\dfrac{a_n}{b_n} = \dfrac{n^3(n^2+1)}{n^5+n+1} \to 1$ as $n \to \infty$. So, by the comparison test, $\sum a_n$ converges since $\sum b_n$ does.

(b)   Let $a_n = n/(3n^2-1)$ and $b_n = 1/n$.   Then

$\dfrac{a_n}{b_n} = \dfrac{n^2}{3n^2-1} \to \dfrac{1}{3}$ as $n \to \infty$.   So, by the comparison test,

$\sum a_n$ diverges since $\sum b_n$ does.

The above example illustrates the working rule that if $a_n = p(n)/q(n)$ where p and q are polynomials of degrees r and s respectively, then it's a good idea to take $b_n = 1/n^{s-r}$ in the comparison test.

EXAMPLE.   Show that the following series are convergent, divergent and divergent respectively:

(a) $\sum \dfrac{\sqrt{n}}{2n^3-1}$,   (b) $\sum \dfrac{n^2+1}{n^2+3n-5}$,   (c) $\sum \dfrac{2n+(-1)^n}{n^2-n+1}$.

151.   THE RATIO TEST

This attempts to settle the convergence/divergence question by looking at the ratio of consecutive terms of the given series.

RESULT.   (Ratio test)   Let a_n ($n \epsilon N$) be positive and suppose that $a_{n+1}/a_n \to L$ as $n \to \infty$.   Then

(i)   if $0 \leq L < 1$, $\sum a_n$ converges,
(ii)   if $L > 1$ or if $a_{n+1}/a_n \to \infty$, then $\sum a_n$ diverges,
(iii)   if $L = 1$, the ratio test fails, but try in turn the techniques of §152 and §276.

Proof.   (i)   Suppose that $0 \leq L < 1$.   Take a number k with $L < k < 1$.   Then find N such that $a_{n+1}/a_n < k$ for all $n \geq N$.   (This is possible because the ratio approaches L as $n \to \infty$.)   Also, by suitably altering the terms $a_1, a_2, \ldots , a_N$ we can ensure that in fact $a_{n+1}/a_n < k$ for all $n \geq 1$.   (These alterations will by §141 not alter the convergent or divergent character of the series.)   It then follows that $a_2 < ka_1$, $a_3 < ka_2$ and so $a_3 < k^2 a_1$, and in general $a_n < k^{n-1} a_1$.   Convergence then follows by comparison with the geometric series $\sum k^{n-1} a_1$, which converges since $0 < k < 1$.

(ii) Proceed in a way similar to part (i), taking k > 1 and showing that $a_n > k^{n-1}a_1$, so that divergence follows by comparison.

N.B.   If you are faced with a particular series you will have to decide whether to apply the comparison test or the ratio test first.   As a rule, if the general term involves a factorial or an nth power it is usually good policy to apply the ratio test rather than the comparison test.

EXAMPLE.   Settle whether the following are convergent:

$$\text{(a) } \sum \frac{n^2}{6^n} \,, \qquad \text{(b) } \sum \frac{(n!)^2}{(2n!)} 5^n \,.$$

Solution.   (a) $\dfrac{a_{n+1}}{a_n} = \dfrac{6^n (n+1)^2}{6^{n+1} n^2} = \dfrac{1}{6}(1 + \dfrac{1}{n})^2 \to \dfrac{1}{6}$ as $n \to \infty$.

So, the series (a) converges by the ratio test as $\frac{1}{6} < 1$.

(b) $\dfrac{a_{n+1}}{a_n} = \dfrac{5(n+1)^2}{(2n+2)(2n+1)} = \dfrac{5(n+1)}{4(n+\frac{1}{2})} \to \dfrac{5}{4}$ as $n \to \infty$.

So, the series (b) diverges by the ratio test as $\frac{5}{4} > 1$.

EXAMPLE.   Prove that the following are all convergent:

$$\text{(a) } \sum \frac{n^7}{6^n} \,, \qquad \text{(b) } \sum \frac{(n+1)^2}{n!} \,, \qquad \text{(c) } \sum \frac{2.5.8\ldots(3n-1)}{4.8.12\ldots(4n)} \,.$$

152.   Gauss's test (§276) has great success in dealing with series for which L = 1 in the ratio test. However, notice the following more elementary technique which has success on some occasions:

Suppose that $a_n \geq 0$ and that $a_{n+1}/a_n \to 1$ as $n \to \infty$, so that the ratio test fails.   If it happens that $a_{n+1}/a_n \geq 1$ for all n > X, then it follows that $a_{n+1} \geq a_n$ for all n > X.   But then $\{a_n\}$ is <u>increasing</u> for n > X and so $a_n \not\to 0$ as $n \to \infty$.   So $\sum a_n$ is divergent by the check of §143.

EXAMPLE.   Test the convergence of $\sum \frac{(n!)^2}{(2n!)} 4^n$.

Solution.   Here $a_{n+1}/a_n = (n+1)/(n+\frac{1}{2}) \to 1$ as $n \to \infty$, so

that the ratio test draws a blank.   However,
$a_{n+1}/a_n > 1$ for all $n\epsilon N$.   So, by the above remark,
$a_n \not\to 0$ as $n\to\infty$.   So $\sum a_n$ is divergent.

N.B.   This technique works only when $a_{n+1}/a_n \geq 1$.
There is no analogue when $a_{n+1}/a_n < 1$.   Series with
$a_{n+1}/a_n < 1$ can be either convergent or divergent: look
at $\sum 1/n^2$ and $\sum 1/n$.

153.   WHICH TEST TO USE ON A SERIES OF NON-NEGATIVE TERMS

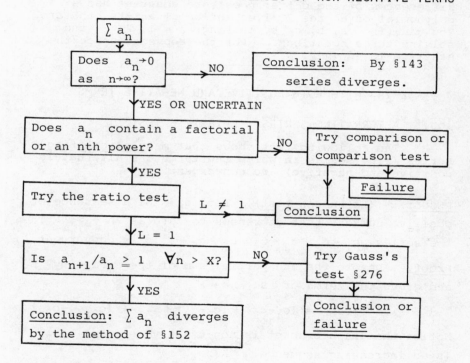

N.B.   This scheme applies only to series with non-
negative terms and even then it is not always successful.
Notice also the integral test ($\S182$) and Cauchy's root
test (Ex.6.22).

154.   THE FORM OF THE NTH TERM

The nth term of a sequence like 7, 11, 15, 19, ... ,
in which the difference between terms is constant, is of
the form (an+b).   Here a is the constant difference
between terms, in this case a = 4, and b is the term

which would naturally precede the first term, in this case b = 3. So, the nth term of the sequence 7, 11, 15, 19, ... is (4n+3).

EXAMPLE. Find the nth term of each of the following:

(a) 4, 10, 16, 22,... (b) -3, 4, 11, 18,...

(c) $\frac{1}{5}$, $\frac{7}{11}$, $\frac{13}{17}$, $\frac{19}{23}$,... (d) $\frac{x}{1.3}$, $\frac{-x^2}{4.6}$, $\frac{x^3}{7.9}$, $\frac{-x^4}{10.12}$,... .

A sequence like 5, 9, 15, 23, 33, ... for which not the first but the second differences (i.e. the differences of the differences) are constant has a polynomial of degree 2 (i.e. an^2+bn+c) as its nth term. You can find a, b, c by letting n = 1, 2, 3 and solving three equations. For the above sequence the nth term is n^2+n+3.

SERIES WITH BOTH POSITIVE AND NEGATIVE TERMS

155. ALTERNATING SERIES: LEIBNIZ'S TEST

The following test shows that many alternating series (i.e. series in which the terms are alternately positive and negative) are convergent.

RESULT. (Leibniz's test) Let $\{c_n\}$ be a sequence of positive numbers that decreases to zero as $n \to \infty$. Then the series $c_1 - c_2 + c_3 - c_4 + c_5 - c_6 + \ldots$ is convergent.

Proof. Denote the nth partial sum of the series by s_n and group the terms of s_{2n} as

$$s_{2n} = (c_1-c_2) + (c_3-c_4) + \ldots + (c_{2n-1}-c_{2n}).$$

Notice that each bracket is non-negative so that $\{s_{2n}\}$ is an increasing sequence. Also,

$$s_{2n} \leq s_{2n} + c_{2n+1} = c_1 - (c_2-c_3) - (c_4-c_5) - \ldots - (c_{2n}-c_{2n+1})$$

$$\leq c_1 .$$

So $\{s_{2n}\}$ is increasing and bounded above and so $s_{2n} \to L$ as $n \to \infty$.

In a similar way, $\{s_{2n+1}\}$ tends to a limit M. Then, $M - L = \lim (s_{2n+1}-s_{2n}) = \lim c_{2n+1} = 0$.

So, L = M, $s_n \to L$ and the series converges to L.

Notice: (i) In the proof of the result it is
actually clear that $c_1 - c_2 \leq s_{2n} \leq c_1$. It follows
that the sum s of the series satisfies $c_1 - c_2 \leq s \leq c_1$.

(ii) The prototype alternating series is the
logarithmic series $1 - \frac{1}{2} + \frac{1}{3} - \frac{1}{4} + \frac{1}{5} - \ldots$, for which
remark (i) gives a sum between ½ and 1. The sum is
actually 0.6931... as is shown in §181.

(iii) A sequence which tends to zero but which does
not decrease does not satisfy the conditions of the test,
e.g. if $c_n = 1/n$ (n odd) and $c_n = 1/n^2$ (n even).

In fact, $1 - \frac{1}{2^2} + \frac{1}{3} - \frac{1}{4^2} + \frac{1}{5} - \frac{1}{6^2} + \ldots$ diverges.

EXAMPLE. Test the following series for convergence:

(a) $\sum \frac{(-1)^n}{\sqrt{(n+1)}\,\sqrt{(n+4)}}$, (b) $\sum \frac{(-1)^{n-1}\,n^3}{n^4+1}$.

Solution. (Notice first that the ratio and comparison
tests cannot be applied because the terms vary in sign.)

(a) Here the terms alternate in sign, and the factors
$\sqrt{(n+1)}$ and $\sqrt{(n+4)}$ increase with n and tend to
infinity as $n \to \infty$. So the sequence $\{|a_n|\}$ decreases
to zero, and we conclude that (a) converges by Leibniz.

(b) Here again the terms alternate in sign and it is
an easy exercise in limits to see that $|a_n| \to 0$ as
$n \to \infty$. To see that the sequence $\{|a_n|\}$ is eventually
decreasing is not so easy; we can do it by showing that
the derivative of the real function f defined by
$f(x) = x^3/(x^4+1)$ is eventually negative as $x \to \infty$.
In fact $f'(x) = (3x^2 - x^6)/(x^4+1)^2 < 0$ for all $x > X$.
(Compare §137.) Leibniz's test then applies and the
series (b) is convergent.

EXAMPLE. Show that the following are convergent,
divergent and convergent respectively:

(a) $\sum (-1)^n \frac{\sqrt{n}}{n+1}$, (b) $\sum \frac{(-1)^n\,n^2}{n^2+1}$, (c) $\sum (-1)^n \sin \frac{1}{n}$.

156. AN IMPORTANT RESULT

The following result relates the behaviour of a real series $\sum a_n$ to that of $\sum |a_n|$. The value of this is that, since $\sum |a_n|$ has non-negative terms, the tests of §149-§152 apply to it.

__RESULT__. Let $a_n \varepsilon R$ $(n \varepsilon N)$. Then if $\sum |a_n|$ converges, so also does $\sum a_n$.

__First proof__. Let $s_n = \sum\limits_{r=1}^{n} a_r$ and $t_n = \sum\limits_{r=1}^{n} |a_r|$.

Notice then from the triangle inequality that

$$|s_n - s_m| \leq |t_n - t_m| \quad \text{for all } m,n \varepsilon N. \quad \ldots (1)$$

Now, since $\sum |a_n|$ converges, $\{t_n\}$ is a convergent sequence and so is Cauchy by §115, i.e.

$$\forall \varepsilon > 0, \exists X \text{ such that } |t_n - t_m| < \varepsilon \quad \forall m,n > X.$$

It then follows from (1) that $|s_n - s_m| < \varepsilon$ $\forall m,n > X$. So, $\{s_n\}$ is a real Cauchy sequence, which must by §116 converge. So $\sum a_n$ converges.

__Second proof__. Define y_n and z_n by

$$y_n = \tfrac{1}{2}(|a_n| + a_n) \quad \text{and} \quad z_n = \tfrac{1}{2}(|a_n| - a_n).$$

Then $0 \leq y_n \leq |a_n|$ and $0 \leq z_n \leq |a_n|$. So $\sum y_n$ and $\sum z_n$ both converge by comparison with $\sum |a_n|$ as in §149. Then $\sum (y_n - z_n)$ converges by §141(iii), i.e. $\sum a_n$ converges as required.

__EXAMPLE__. Test the series $\sum (-1)^n \dfrac{\sin\sqrt{n}\pi}{n^2}$ for convergence.

__Solution__. Here $|a_n| = \dfrac{|\sin\sqrt{n}\pi|}{n^2} \leq \dfrac{1}{n^2}$ for all $n \varepsilon N$. So, by comparison (§149) with $\sum 1/n^2$, deduce that $\sum |a_n|$ converges. So, by the above result, $\sum a_n$ converges.

(Notice that Leibniz's test is of no use here since the terms are _not_ alternating though superficially so.)

Look also at Ex.6.23 at the end of the chapter.

157. ABSOLUTE CONVERGENCE

Within the set of all convergent series we distinguish a special subset called <u>absolutely convergent</u> <u>series</u>. $\sum a_n$ qualifies for membership of this subset (i.e. is <u>absolutely convergent</u>) provided that $\sum |a_n|$ is convergent. (The results of §156 and §168 guarantee that the set of absolutely convergent series <u>really is</u> a subset of the set of all convergent series.) All other convergent series that fail to qualify for membership of this subset (i.e. that are not absolutely convergent) are called <u>conditionally convergent</u>. So the set of all series is split into three disjoint subsets thus:

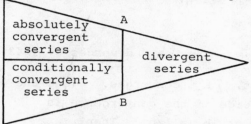

N.B. The set of all convergent series is the union of the two disjoint subsets on the left of line AB.

As examples notice that $1 - \frac{1}{2^2} + \frac{1}{3^2} - \frac{1}{4^2} + \frac{1}{5^2} - \frac{1}{6^2} + \ldots$ <u>is</u> absolutely convergent because $1 + \frac{1}{2^2} + \frac{1}{3^2} + \frac{1}{4^2} + \ldots$ is convergent, whereas $1 - 1 + \frac{1}{2} - \frac{1}{2} + \frac{1}{3} - \frac{1}{3} + \ldots$ is only conditionally convergent because $1 + 1 + \frac{1}{2} + \frac{1}{2} + \frac{1}{3} + \frac{1}{3} + \ldots$ is divergent.

158.

To settle the convergence/divergence question for a series $\sum a_n$ of real terms, first see if Leibniz's test will apply. If not, take $\sum |a_n|$ and try to decide if it is convergent using the scheme of §153. If $\sum |a_n|$ <u>is</u> convergent, then so also is $\sum a_n$ by §156.

Be warned however that if $\sum |a_n|$ is divergent, there is no guarantee that $\sum a_n$ diverges: for example, look at any conditionally convergent series, like $1 - 1 + \frac{1}{2} - \frac{1}{2} + \frac{1}{3} - \frac{1}{3} + \ldots$. In cases like this where $\sum |a_n|$ turns out to be divergent we must resort to other methods to decide convergence or divergence for $\sum a_n$.

159. AN EXAMPLE ON ABSOLUTE AND CONDITIONAL CONVERGENCE

EXAMPLE. Decide whether the series $\sum a_n$ is absolutely convergent, conditionally convergent or divergent when a_n has each of the following forms:

(a) $\dfrac{\sin n}{n^2}$, (b) $\dfrac{(-1)^n n^4}{n^4+1}$, (c) $\dfrac{(-1)^n n^3}{n^4+1}$, (d) $\dfrac{(-1)^n n^2}{n^4+1}$.

Solution. (a) $\left| \dfrac{\sin n}{n^2} \right| \leqq \dfrac{1}{n^2}$ for all $n \in N$.

Then since $\sum \dfrac{1}{n^2}$ converges, $\sum \dfrac{\sin n}{n^2}$ is absolutely convergent by comparison.

(b) $a_n \not\to 0$ as $n \to \infty$. So $\sum a_n$ diverges by §143.

(c) By Leibniz's test, $\sum a_n$ converges. (See the example in §155.) To decide if the convergence is absolute or conditional, let $b_n = 1/n$. Then,

$$\frac{|a_n|}{b_n} = \frac{n^4}{n^4+1} \to 1 \quad \text{as} \quad n \to \infty,$$

so that $\sum |a_n|$ diverges by the comparison test because $\sum 1/n$ diverges. So $\sum a_n$ is conditionally convergent.

(d) As in (c), $\sum a_n$ converges by Leibniz's test. To decide if the convergence is absolute or conditional, let $b_n = 1/n^2$. Then,

$$\frac{|a_n|}{b_n} = \frac{n^4}{n^4+1} \to 1 \quad \text{as} \quad n \to \infty,$$

so that $\sum |a_n|$ converges by the comparison test because $\sum 1/n^2$ converges. So $\sum a_n$ is absolutely convergent.

160. EXAMPLE. Discuss the convergence of $\sum \dfrac{n^2 x^n}{(2n+1)^3}$ for all values of $x \in R$.

Solution. Here $|a_n| = \dfrac{n^2 |x|^n}{(2n+1)^3}$. Apply the ratio test because there is an nth power present.

$$\frac{|a_{n+1}|}{|a_n|} = \frac{(2n+1)^3}{(2n+3)^3} \cdot \frac{(n+1)^2}{n^2} |x| \to |x| \quad \text{as} \quad n \to \infty.$$

So $\sum |a_n|$ converges for $|x| < 1$ and so $\sum a_n$ converges absolutely for $|x| < 1$.

For $x = 1$, $a_n = n^2/(2n+1)^3$. This diverges by comparing with $\sum 1/n$ in the comparison test.

For $x = -1$, $a_n = (-1)^n n^2/(2n+1)^3$ and this series converges by Leibniz's test.

For $|x| > 1$, $|a_n| = n^2|x|^n/(2n+1)^3$ and this tends to infinity as $n \to \infty$ by §63. So, in this case $|a_n| \not\to 0$ as $n \to \infty$ and so the series diverges for all x with $|x| > 1$.

So, the overall conclusion is that $\sum a_n$ converges for $x \in [-1,1[$ and diverges for all other $x \in R$.

(Notice that the results on radius of convergence of §190 and §191 remove the need to check what happens for $|x| > 1$ once the initial application of the ratio test has been made.)

161. **EXAMPLE.** Test the convergence of $\sum (\sqrt{(n+1)} - \sqrt{n})^k$ for all values of $k \in R$.

Solution. Though $\sqrt{(n+1)}$ and $\sqrt{n}$ are large when n is large, the difference between them is small. To clarify the situation notice that

$$\sqrt{(n+1)} - \sqrt{n} = \frac{(\sqrt{(n+1)} - \sqrt{n})(\sqrt{(n+1)} + \sqrt{n})}{(\sqrt{(n+1)} + \sqrt{n})} = \frac{1}{(\sqrt{(n+1)} + \sqrt{n})}.$$

So, for the given series, $a_n = \dfrac{1}{(\sqrt{(n+1)} + \sqrt{n})^k}$, and here the denominator behaves like $(2\sqrt{n})^k$ for large values of n. So let $b_n = 1/n^{\frac{1}{2}k}$ in the comparison test and find that $a_n/b_n \to 1/2^k$ as $n \to \infty$. Conclude from this that $\sum a_n$ converges or diverges according as $\sum b_n$ does. So $\sum a_n$ converges for $k > 2$ and diverges for $k \le 2$.

EXAMPLE. Show the following are convergent and divergent respectively: (For (b) look at §134(ii).)

(a) $\sum (\sqrt{(n^3+1)} - \sqrt{(n^3-1)})$, (b) $\sum (\log(n+1) - \log n)^{\frac{1}{2}}$.

SERIES WITH COMPLEX TERMS

162. PROPERTIES OF COMPLEX NUMBERS

Take the field of complex numbers as known. Let $z = x+iy$ $(x, y \epsilon R)$ with $i^2 = -1$. Then $\bar{z} = x-iy$ and $|z| = \sqrt{(x^2+y^2)}$. Notice the following properties (roughly in increasing order of usefulness):

$$Re(z+w) = Re\ z + Re\ w, \qquad Im(z+w) = Im\ z + Im\ w;$$

$$|\bar{z}| = |z|, \quad z + \bar{z} = 2Re\ z, \quad z - \bar{z} = 2iIm\ z;$$

$$|Re\ z| \leq |z|, \qquad |Im\ z| \leq |z|;$$

$$\overline{z + w} = \bar{z} + \bar{w}, \qquad \overline{zw} = \bar{z}\bar{w};$$

$$z.\bar{z} = |z|^2.$$

163.

(i) <u>RESULT</u>. $|ab| = |a||b|$ for all $a, b \epsilon C$.

<u>Proof</u>. $|ab|^2 = ab.\overline{ab} = ab.\bar{a}\bar{b} = a\bar{a}b\bar{b} = |a|^2|b|^2$. (It follows further that $|a^n| = |a|^n$ for $n \epsilon N$.)

(ii) <u>RESULT</u>. (<u>The triangle inequality</u>) For $a, b \epsilon C$,

$$|a + b| \leq |a| + |b|.$$

<u>Proof</u>.
$$\begin{aligned}
|a + b|^2 &= (a + b)\overline{(a + b)} = (a + b)(\bar{a} + \bar{b}) \\
&= a\bar{a} + b\bar{b} + (a\bar{b} + \bar{a}b) \\
&= |a|^2 + |b|^2 + 2Re\ a\bar{b} \\
&\leq |a|^2 + |b|^2 + 2|a\bar{b}| \\
&= |a|^2 + |b|^2 + 2|a||b| \\
&= (|a| + |b|)^2.
\end{aligned}$$

Hence the result.

(iii) <u>RESULT</u>. For $a, b \epsilon C$, $||a| - |b|| \leq |a - b|$.

<u>Proof</u>. Deduce this from (ii) by analogy with the proof of its analogue in §9.

164. C AS A METRIC SPACE: CONVERGENCE OF A COMPLEX SEQUENCE

C with $d(z,w) = |z - w|$ is a metric space (§94).

M1 and M2 are clear. M3 follows from the triangle inequality of §163 since

$$|z_1 - z_3| \leq |z_1 - z_2| + |z_2 - z_3|,$$
$$\text{i.e. } d(z_1,z_3) \leq d(z_1,z_2) + d(z_2,z_3).$$

The <u>open neighbourhood</u> $N(z,\varepsilon)$ is an open disc of radius ε centred at the point z. The sequence $\{z_n\}$ <u>converges</u> to z means:

$$\forall \varepsilon > 0, \exists X \text{ such that } z_n \varepsilon N(z,\varepsilon) \quad \forall n > X,$$

i.e. $\forall \varepsilon > 0, \exists X$ such that $|z_n - z| < \varepsilon \ \forall n > X.$

Notice the following two results:

<u>RESULT</u>. $\{z_n\}$ converges to z as $n \to \infty$ if and only if $\text{Re } z_n \to \text{Re } z$ and $\text{Im } z_n \to \text{Im } z$ as $n \to \infty$.

<u>Proof</u>. Let $z_n = x_n + iy_n$ and $z = x + iy$ $(x_n, y_n, x, y \varepsilon R)$.

Since $|x_n - x| = |\text{Re}(z_n - z)| \leq |z_n - z|$

conclude that if $z_n \to z$ then $x_n \to x$ and similarly $y_n \to y$.

Also,

$$|z_n - z| = |x_n + iy_n - x - iy| \leq |x_n - x| + |y_n - y|.$$

So if $x_n \to x$ and $y_n \to y$ it follows that $z_n \to z$.

<u>RESULT</u>. If $z_n \to z$ as $n \to \infty$, then $|z_n| \to |z|$ as $n \to \infty$.

<u>Proof</u>. By §163, $||z_n| - |z|| \leq |z_n - z|$. Hence the result.

N.B. These two results are instrumental in reducing the problem of the convergence of many complex sequences to the case of <u>real</u> sequences.

165. There is no natural order for complex numbers, i.e. $z_1 < z_2$ is meaningless. For this reason we cannot talk of a complex sequence tending to infinity.

166. COMPLEX SEQUENCES THAT CONVERGE TO ZERO

The complex sequence $\{a_n\}$ converges to zero if and only if the real sequence $\{|a_n|\}$ converges to zero. This is because the main clause in each definition reduces to $|a_n| < \varepsilon$.

Consequently notice that if $|z| < 1$ then $z^n \to 0$ as $n \to \infty$, since $|z^n| = |z|^n$. Notice also that if $|z| \geq 1$ then $|z^n| = |z|^n \geq 1$ and $z^n \not\to 0$.

EXAMPLE. By taking the modulus prove that if $|z| < 1$ and $k \varepsilon R$, then $n^k z^n \to 0$ as $n \to \infty$.

EXAMPLE. By taking the modulus prove that for all $z \varepsilon C$ $z^n/n! \to 0$ as $n \to \infty$. (Compare the method of §65.)

167. THE COMPLEX GEOMETRIC SERIES

This is $1+z+z^2+z^3+ \ldots$ $(z \varepsilon C)$. Using the result about z^n from §166 together with the partial sums of §144 we see that the series converges only for $|z| < 1$.

168. CONVERGENCE OF A GENERAL COMPLEX SERIES

This can often be settled by considering the series of non-negative terms $\sum |a_n|$ (to which the scheme of §153 may be applied) and then using the following result, which generalises the result of §156.

RESULT. Let $\sum a_n$ be a complex series. Then if $\sum |a_n|$ converges so also does $\sum a_n$.

First proof. The first proof of the corresponding result for real series in §156 carries over verbatim, except that $\{s_n\}$ is a complex Cauchy sequence and this converges by §117 not §116.

Second proof. Suppose that $\sum |a_n|$ converges. Then

$0 \leq |\text{Re } a_n| \leq |a_n|$ and $0 \leq |\text{Im } a_n| \leq |a_n|$. So, use comparison(§149) to deduce that $\sum |\text{Re } a_n|$ and $\sum |\text{Im } a_n|$ converge. Then use the result of §156 to conclude that $\sum \text{Re } a_n$ and $\sum \text{Im } a_n$ both converge. Then $\sum (\text{Re } a_n + i\text{Im } a_n)$ converges, i.e. $\sum a_n$ converges.

169. Notice that the definitions of absolute and conditional convergence of §157 apply to <u>complex</u> series as well as to real ones. A convergent complex series falls into one of the two disjoint subsets, the absolutely convergent series or the conditionally convergent series according as $\sum |a_n|$ converges or diverges respectively.

Recall from §158, (and this applies to complex series as well as to real ones), that if $\sum |a_n|$ diverges then $\sum a_n$ can still be either conditionally convergent or divergent. Notice how we decide between these possibilities when $|z| > 1$ in §170.

170. <u>EXAMPLE</u>. Discuss the convergence of $\sum z^n/n^2$ for all values of $z \varepsilon C$.

<u>Solution</u>. Apply the ratio test to $\sum |a_n|$ since there is an nth power present. Then

$$\frac{|a_{n+1}|}{|a_n|} = \frac{|z| n^2}{(n+1)^2} \to |z| \quad \text{as} \quad n \to \infty.$$

So $\sum a_n$ is absolutely convergent for $|z| < 1$.

When $|z| = 1$, $|a_n| = 1/n^2$ and $\sum 1/n^2$ converges. So $\sum a_n$ is absolutely convergent for $|z| = 1$ too.

When $|z| > 1$, $|a_n| = |z|^n/n^2$ and this tends to infinity as $n \to \infty$ by §63(i) with the result that the series diverges by §143.

Notice: (i) The behaviour of the above series — convergent for $|z| \leq 1$ and divergent for $|z| > 1$ is part of the general picture of the circle of convergence of §190 and §191. The check on what happens for $|z| > 1$ is rendered unnecessary after the initial application of the ratio test by these later results.

(ii) The behaviour of $\sum z^n/n$ for $|z| = 1$ is much more difficult than the above. See §275.

EXAMPLES 6

1. Settle whether the series $\sum a_n$ is convergent, where a_n has each of the following forms:

(a) $\dfrac{n}{n^2+1}$, (b) $\dfrac{n^2+2}{n^4+n^2+1}$, (c) $\dfrac{n}{3n+1}$, (d) $\dfrac{3^n}{n^2}$,

(e) $\dfrac{n^2}{3^n}$, (f) $\dfrac{n^2+1}{3^n+1}$, (g) $\dfrac{2^n}{n!}$, (h) $\dfrac{n^2}{2n^2+n+1}$,

(i) $\cos^2\left(\dfrac{1}{n}\right)$, (j) $1/(\sqrt{n}+\sqrt{(n+1)})$, (k) $(\sqrt{(n+1)}-\sqrt{n})^3$,

(l) $\dfrac{(n!)^2}{(2n)!}$, (m) $\dbinom{2n}{n}(\tfrac{1}{2})^n$, (n) $\dfrac{1}{n^{1+1/n}}$, (o) $\dfrac{n^{10}}{n!}$,

(p) $\dfrac{2.6.10\ldots(4n-2)}{5.8.11\ldots(3n+2)}\cdot\left(\dfrac{2}{3}\right)^n$, (q) $\dfrac{5.8.11\ldots(3n+2)}{6.10.14\ldots(4n+2)}\cdot\left(\dfrac{4}{3}\right)^n$,

(r) $n!/3^{n^2}$, (s) $(n!)^3\,27^n/(3n)!$, (t) $(n+1)^2/n^n$,

(u) $1/n^{\sqrt{n}}$.

2. Show that $\sum \dfrac{p(n)}{q(n)}$, where p and q are polynomials of degrees α and β respectively, is convergent if and only if $\beta > \alpha+1$.

3. Prove that if $\sum a_n$ converges and $\sum b_n$ diverges, then $\sum (a_n+b_n)$ diverges. [Hint: Use a contradiction.]

4. Use the properties of geometric series to express $0.\overset{\bullet}{3}\overset{\bullet}{6}$ (i.e. $0.363636\ldots$) and $0.3\overset{\bullet}{1}0\overset{\bullet}{8}$ as vulgar fractions.

5. Let $s_n = 1 + 2x + 3x^2 + \ldots + nx^{n-1}$. Evaluate $s_n - xs_n$ and deduce that, for $-1 < x < 1$, $s_n \to (1-x)^{-2}$ as $n \to \infty$. (Compare Ex.7.6 and §198.)

6. Show that $\sum \left(\dfrac{a}{n} + \dfrac{b}{n+1} - \dfrac{c}{n+2}\right)$ converges if and only if $c = a + b$.

7. For $n \geq 2$, show that $\dfrac{2}{n(n^2-1)} = \dfrac{1}{n-1} - \dfrac{2}{n} + \dfrac{1}{n+1}$. Hence find the sum of $\dfrac{1}{1.2.3} + \dfrac{1}{2.3.4} + \dfrac{1}{3.4.5} + \ldots$.

8. Settle whether the series $\sum a_n$ is absolutely convergent, conditionally convergent or divergent for each of the following forms of a_n:

(a) $\dfrac{(-1)^n n}{n^2+1}$,   (b) $\dfrac{(-1)^n \sin n}{3^n}$,

(c) $\dfrac{(-1)^n \sqrt{n}}{\sqrt{(n+1)}}$,   (d) $\dfrac{(-1)^n n^2}{2^n (n+1)}$.

9. For each of the following determine the range of real values of x for which the series converges:

(a) $\sum \dfrac{n^2}{n+1}\left(\dfrac{x}{3}\right)^n$,   (b) $\sum \dfrac{n! x^n}{(2n)!}$,

(c) $\sum \dfrac{(-1)^n x^n}{2^n \sqrt{n}}$,   (d) $\sum n! x^n$.

10. Guess the values of $k > 0$ for which the series $\sum \dfrac{(-1)^n n^k}{n^2+1}$ is absolutely/conditionally convergent. Prove that your guess is correct.

11. By doing <u>suitable</u> estimation, prove that the following are all divergent:

(a) $1 + \dfrac{1}{6} + \dfrac{1}{11} + \dfrac{1}{16} + \dots$,

(b) $1 - \dfrac{1}{2} + \dfrac{1}{3} - \dfrac{1}{4} + \dfrac{1}{5} + \dfrac{1}{6} - \dfrac{1}{7} + \dfrac{1}{8} - \dfrac{1}{9} + \dfrac{1}{10} + \dfrac{1}{11} - \dots$,

(c) $1 - \dfrac{1}{2} - \dfrac{1}{3} + \dfrac{1}{4} + \dfrac{1}{5} + \dfrac{1}{6} - \dfrac{1}{7} - \dfrac{1}{8} + \dfrac{1}{9} + \dfrac{1}{10} + \dfrac{1}{11} - \dots$.

[N.B. In (b) and (c) the pattern of signs repeats every five terms.]

12. If $\sum_{n=1}^{\infty} 1/(n^2+a^2)^2 = K$ and $\sum_{n=1}^{\infty} n^2/(n^2+a^2)^2 = L$, find $\sum_{n=1}^{\infty} 1/(n^2+a^2)$ in terms of K and L.

13. In this question all the series are real. Give examples to show that

(a) if $a_n > 0$ and $\sum a_n^2$ converges, $\sum a_n$ can diverge,

(b) if $a_n > 0$ and $a_{n+1}/a_n < 1$ for all $n > X$,

$\sum a_n$ can still diverge,

(c) if $a_n > 0$, $b_n > 0$ and $a_n/b_n \to 0$ as $n \to \infty$, $\sum a_n$ and $\sum b_n$ need not both converge or both diverge,

(d) if $\sum a_n$ converges and $a_n > 0$ $(n \in N)$, it does not follow that there exists X such that $a_{n+1}/a_n \leq 1$ for all $n > X$,

(e) if $\sum a_n$ converges and $b_n \to 0$ as $n \to \infty$, then $\sum a_n b_n$ can diverge,

(f) if $\sum a_n$ converges, $\sum a_n^2$ may diverge while $\sum a_n^3$ converges,

(g) if $a_n > 0$ $(n \in N)$ and $\sum a_n$ converges, it is possible that $\sum n^k a_n$ converges for every $k \in N$,

(h) if $a_n \geq 0$ and $b_n \geq 0$ $(n \in N)$ and $\sum \sqrt{(a_n b_n)}$ converges, both $\sum a_n$ and $\sum b_n$ can diverge,

(i) if $a_n > 0$ $(n \in N)$ and $na_n \to 0$ as $n \to \infty$, $\sum a_n$ can diverge, (See §215.)

(j) if $a_n \geq 0$ and $(a_{n+1} + a_{n+2} + \ldots + a_{2n}) \to 0$ as $n \to \infty$, $\sum a_n$ can diverge,

(k) if $a_n > 0$ and $\sum a_n$ converges, then na_{n^2} need not tend to zero as $n \to \infty$.

14. A positive integer is a _palindrome_ if it reads the same forwards as backwards (e.g. 3553). The numbers of palindromes in the intervals [1,9], [10,99], [100,999] are 9, 9 and 90. Prove a general result of this type and by doing estimation, show that $\sum 1/k$ where k runs over all palindromes converges to a sum less than 11.

15. Prove the following results:
(i) if $a_n \geq 0$ $(n \in N)$ and $\sum a_n$ converges, $\sum a_n^2$ converges,

(ii) if $a_n \geq 0$, $b_n \geq 0$ $(n \in N)$ and $\sum a_n$ and $\sum b_n$ converge, then $\sum \sqrt{(a_n b_n)}$ converges, (Use Cauchy's inequality from §277.)

(iii) if $\sum a_n$ is absolutely convergent and $b_n \to L$ as $n \to \infty$, then $\sum a_n b_n$ is absolutely convergent.

16. By splitting the series $\sum i^n/n$ into its real and imaginary parts, show that it converges. (See Ex.10.8.)

17. Prove that for all complex numbers a and b,
(i) $|a+b|^2 + |a-b|^2 = 2(|a|^2 + |b|^2)$,
(ii) $4ab = |b+\bar{a}|^2 - |b-\bar{a}|^2 + i|b+i\bar{a}|^2 - i|b-i\bar{a}|^2$.

18. Find all values of $z \varepsilon C$ such that the sequence $\{z^{4n}\}$ converges as $n \to \infty$.

19. $\sum |a_n|$ diverges, where $a_n = b_n + ic_n$ and b_n, c_n are non-negative. Prove that at least one of $\sum b_n$ and $\sum c_n$ must diverge.

20. Show that if $\sum a_n$ and $\sum b_n$ are absolutely convergent, then $\sum a_n \cos nx$ and $\sum b_n \sin nx$ converge for all $x \varepsilon R$. Suppose that f is an _even_ function and that
$$f(x) = \sum_{n=1}^{\infty} (a_n \cos nx + b_n \sin nx) \quad \text{for all } x \varepsilon R.$$
Prove that $f(x) = \sum_{n=1}^{\infty} a_n \cos nx$ for all $x \varepsilon R$.

21. In the series $1 - \frac{1}{2} - \frac{1}{3} + \frac{1}{4} + \frac{1}{5} + \frac{1}{6} + \frac{1}{7} - \frac{1}{8} - ..$ successive blocks of 1, 2, 4, 8, ... terms have the same sign. Prove that this series diverges.

22. (<u>Cauchy's root test</u>) Let $a_n > 0$ and suppose that $a_n^{1/n} \to L$ as $n \to \infty$. Then
(i) if $0 \leq L < 1$, $\sum a_n$ converges,
(ii) if $L > 1$ or if $a_n^{1/n} \to \infty$, $\sum a_n$ diverges,
(iii) if $L = 1$, the test is inconclusive.

Sketch a proof of the test and use it to discuss the convergence of (a) $\sum \left(\frac{n}{2n+1}\right)^n$, (b) $\sum 3^n 5^{-n^2}$,

E

(c) $\sum (\log n)^{-n}$.   (The result of Ex.13.3 shows that if the ratio test gives a conclusion then so also will the root test.   Moreover there are series for which the root test gives a conclusion while the ratio test fails, e.g. the series $a + a + a^2 + a^2 + a^3 + a^3 + \ldots$ with $0 < a < 1$.   However, despite this the ratio test is often the more useful in practice because a_{n+1}/a_n is easier to handle than $a_n^{1/n}$.)

23.   Show by the method used in the example in §156 that $\sum \dfrac{\sin \sqrt{n}\pi}{n^3}$ converges.   Show in contrast that this method gives no conclusion about the convergence of $\sum \dfrac{\sin \sqrt{n}\pi}{n}$.

VII. TAYLOR AND MACLAURIN EXPANSIONS

171. THE BACKGROUND TO TAYLOR'S THEOREM

Suppose that the real function f is many times differentiable on R. Suppose we know $f(a)$ but not $f(b)$. The mean-value theorem (§129) can be written as

$$f(b) = f(a) + (b-a)f'(\xi), \quad \text{where} \quad \xi \varepsilon]a,b[. \quad ...(*)$$

Here think of a as a fixed basic point and of b as a neighbouring point. The statement (*) says that $f(b)$ is $f(a)$ plus an extra piece, but unfortunately we do not know ξ. Taylor's theorem gives a more delicate form of (*), e.g. with four terms,

$$f(b) = f(a) + (b-a)f'(a) + \frac{(b-a)^2}{2!}f''(a) + \frac{(b-a)^3}{3!}f'''(\xi),$$

$$\text{where} \quad \xi \varepsilon]a,b[.$$

Here, if the final term is small, we can get a good estimate of $f(b)$ from the first three terms which use information about f <u>at the point a only</u>.

To see how Taylor's theorem works, look at the following example, which is perhaps more helpful than the proof.

<u>EXAMPLE</u>. Let $f(x) = x^3 + 17x + 38$ $(x \varepsilon R)$. Use the above form of Taylor's theorem to expand f in the cases when (i) $a = 0$, (ii) $a = 1$.

<u>Solution</u>. (i) Take $a = 0$. Then $f'(x) = 3x^2 + 17$, $f''(x) = 6x$ and $f'''(x) = 6$, so that $f(0) = 38$, $f'(0) = 17$, $f''(0) = 0$ and $f'''(\xi) = 6$ for every ξ. So the above expansion gives

$$f(b) = 38 + b.17 + \frac{b^2}{2!}.0 + \frac{b^3}{3!}.6 = 38 + 17b + b^3,$$

as expected.

(ii) Take $a = 1$. Then $f(1) = 56$, $f'(1) = 20$, $f''(1) = 6$ and $f'''(\xi) = 6$ for every ξ. So,

$$f(b) = 56 + (b-1).20 + \frac{(b-1)^2}{2!}.6 + \frac{(b-1)^3}{3!}.6$$

$$= 56 + 20(b-1) + 3(b-1)^2 + (b-1)^3.$$

(Check that this multiplies out to give $b^3 + 17b + 38$.)

172. TAYLOR'S THEOREM

<u>RESULT</u>. Let f be defined on $[a,b]$. Let $f^{(n-1)}$ exist and be continuous on $[a,b]$ and let $f^{(n)}$ exist on $]a,b[$. Then

$$f(b) = f(a) + (b-a)f'(a) + \frac{(b-a)^2}{2!}f''(a) + \ldots$$

$$\ldots + \frac{(b-a)^{n-1}}{(n-1)!}f^{(n-1)}(a) + R_n(b),$$

where $R_n(b) = \frac{(b-a)^n f^{(n)}(\xi)}{n!}$ (Lagrange's remainder),

or $R_n(b) = \frac{(b-a)(b-\xi)^{n-1}}{(n-1)!}f^{(n)}(\xi)$ (Cauchy's remainder),

where ξ lies between a and b. (The same result holds under analogous conditions if $b < a$.)

<u>Proof</u>. Define a continuous function ϕ on $[a,b]$ by

$$\phi(x) = f(b) - f(x) - (b-x)f'(x) - \frac{(b-x)^2}{2!}f''(x) - \ldots$$

$$\ldots - \frac{(b-x)^{n-1}}{(n-1)!}f^{(n-1)}(x).$$

Notice that $\phi(b) = 0$ and that $\phi(a)$ is the difference between $f(b)$ and the non-remainder terms of the would-be expression for $f(b)$. Also ϕ is differentiable on $]a,b[$. All we need prove is that $\phi(a) = R_n(b)$. Notice also that

$$\phi'(x) = -\frac{(b-x)^{n-1}f^{(n)}(x)}{(n-1)!}$$ (because of a collapse).

Now let $g(x) = (b-a)\phi(x) - (b-x)\phi(a)$. Then $g(a) = g(b) = 0$. So, by Rolle's theorem, there exists $\xi \in]a,b[$ such that $g'(\xi) = 0$,

i.e. $(b-a)\phi'(\xi) + \phi(a) = 0$.

So $\phi(a) = -(b-a)\phi'(\xi) = \frac{(b-a)(b-\xi)^{n-1}}{(n-1)!}f^{(n)}(\xi)$,

i.e. Cauchy's remainder. From our comment about $\phi(a)$ above, the result is proved.

To get Lagrange's remainder, instead of looking at g, look at h where $h(x) = (b-a)^n\phi(x) - (b-x)^n\phi(a)$. Here $h(a) = h(b) = 0$ and again Rolle applies and shows that $\phi(a)$ is equal to Lagrange's remainder.

Notice: (i) Essentially the same proof applies if b < a.

(ii) Both remainders are useful, e.g. in §174.

173. MACLAURIN'S THEOREM

Taylor's theorem indicates how to expand a function in terms of its derivatives at a general point a. For example, in §171 we took a = 0 and a = 1. Maclaurin's theorem is the special case of Taylor's theorem given by a = 0: i.e. under suitable conditions,

$$f(x) = f(0) + xf'(0) + \frac{x^2}{2!}f''(0) + \ldots$$

$$\ldots + \frac{x^{n-1}}{(n-1)!}f^{(n-1)}(0) + R_n(x),$$

where $R_n(x) = \frac{x^n}{n!}f^{(n)}(\xi)$ or $\frac{x(x-\xi)^{n-1}}{(n-1)!}f^{(n)}(\xi)$ in the

Lagrange and Cauchy forms. (Notice that x can be positive or negative and that ξ lies between 0 and x.)

The series expansions of many common functions are Maclaurin series, i.e. the infinite series arising from the above if the remainder term tends to zero as $n \to \infty$. Cases in point are the series for sin x, log(1+x), cosh x and 1/(1-x).

174. EXAMPLE. Use Taylor's theorem with a = 0

(i.e. Maclaurin's theorem) to develop a series expansion for f, where f(x) = 1/(1-x).

Solution. It is routine to see that

$$f^{(n)}(x) = \frac{n!}{(1-x)^{n+1}} \quad (n \epsilon N).$$

So f(0) = 1, f'(0) = 1, f''(0) = 2!, ... , $f^{(n)}(0) = n!$.
So, from Taylor's theorem with a = 0,

$$\frac{1}{(1-x)} = 1 + x.1 + \frac{x^2}{2!}.2! + \frac{x^3}{3!}.3! + \ldots$$

$$\ldots + \frac{x^{n-1}}{(n-1)!}.(n-1)! + R_n(x)$$

$$= 1 + x + x^2 + x^3 + \ldots + x^{n-1} + R_n(x).$$

In the Lagrange form $R_n(x) = \frac{x^n}{n!}.\frac{n!}{(1-\xi)^{n+1}}$. If

$x \epsilon]-1,0]$, then $(1-\xi) \geq 1$, so that $|R_n(x)| \leq |x|^n \to 0$

as $n \to \infty$, since $-1 < x \leq 0$. So, for $x \varepsilon]-1,0]$, we can legitimately write

$$\frac{1}{(1-x)} = 1 + x + x^2 + x^3 + \ldots . \qquad \ldots (*)$$

From work on the geometric series (§144) we know that (*) is actually valid for all $x \varepsilon]-1,1[$, but it is not clear that the Lagrange remainder tends to zero for all such x. A slight modification of the above argument also covers the case where $x \varepsilon]0,\frac{1}{2}[$, but for $x \varepsilon [\frac{1}{2},1[$ we are forced to consider the Cauchy remainder. The following such argument covers all $x \varepsilon]0,1[$. Let $x \varepsilon]0,1[$. Then there exists $\xi \varepsilon]0,x[$ such that

$$R_n(x) = \frac{nx(x-\xi)^{n-1}}{(1-\xi)^{n+1}} .$$

So, $|R_n(x)| \leq \dfrac{nx}{(1-x)^2} \cdot \left(\dfrac{x-\xi}{1-\xi}\right)^{n-1} = \dfrac{nx}{(1-x)^2} \cdot \left(\dfrac{x-\xi}{(x-\xi)+(1-x)}\right)^{n-1}$

$$\leq \frac{nx}{(1-x)^2} \cdot \left(\frac{x}{x+(1-x)}\right)^{n-1} \quad \text{(by Ex.1.2, since}$$
$$0 < x-\xi < x < 1)$$

$$\leq \frac{nx^n}{(1-x)^2}$$

$$\to 0 \quad \text{as} \quad n \to \infty, \text{ by §63.}$$

So we have justified the expansion $(1-x)^{-1} = 1+x+x^2+\ldots$ for all $x \varepsilon]-1,1[$.

Notice that if we try to put $x = -2$ in this last example the two remainders give

$$|R_n(-2)| = 2^n/(1-\xi)^{n+1} \quad \text{or} \quad 2n(\xi+2)^{n-1}/(1-\xi)^{n+1}$$

where $\xi \varepsilon]-2,0[$. However there is no guarantee that these tend to zero as $n \to \infty$. In fact $R_n(-2)$ must be massive because the terms $1-2+2^2-2^3+\ldots+(-1)^{n-1}2^{n-1}$ are useless as an approximation to $(1-(-2))^{-1}$, i.e. 1/3.

Taylor's theorem is only of use for those x for which the remainder term $R_n(x)$ tends to zero as $n \to \infty$. See also §224.

175. THE GENERAL BINOMIAL THEOREM

The expansions

$$(1+x)^3 = 1 + 3x + 3x^2 + x^3 \qquad (x \varepsilon R)$$

and
$$(1+x)^{-1} = 1 - x + x^2 - x^3 + \ldots \qquad (-1 < x < 1)$$

are particular parts of one big picture — the general binomial theorem.

For every real number α (including fractions and negative numbers) and for every positive integer r, define the <u>binomial coefficient</u> $\begin{pmatrix} \alpha \\ r \end{pmatrix}$ by

$$\begin{pmatrix} \alpha \\ r \end{pmatrix} = \frac{\alpha(\alpha-1)(\alpha-2) \ldots (\alpha-r+1)}{r!} .$$

Also define $\begin{pmatrix} \alpha \\ 0 \end{pmatrix} = 1$.

The theorem says that, for $-1 < x < 1$,

$$(1+x)^\alpha = \begin{pmatrix} \alpha \\ 0 \end{pmatrix} + \begin{pmatrix} \alpha \\ 1 \end{pmatrix} x + \begin{pmatrix} \alpha \\ 2 \end{pmatrix} x^2 + \begin{pmatrix} \alpha \\ 3 \end{pmatrix} x^3 + \ldots .$$

When α is a positive integer the expansion terminates after $(\alpha+1)$ terms and in that case the result is valid for all $x \varepsilon R$.

<u>RESULT</u>. Let $\alpha \varepsilon R$. Then, for $-1 < x < 1$,

$$(1+x)^\alpha = \sum_{n=0}^{\infty} \begin{pmatrix} \alpha \\ n \end{pmatrix} x^n.$$

<u>Proof.</u> Use Taylor's theorem with $a = 0$. Let $F(x) = (1+x)^\alpha$. Then $F'(x) = \alpha(1+x)^{\alpha-1}$, $F''(x) = \alpha(\alpha-1)(1+x)^{\alpha-2}$ and in general

$$F^{(n)}(x) = \alpha(\alpha-1) \ldots (\alpha-n+1)(1+x)^{\alpha-n}$$

so that $F^{(n)}(0) = \alpha(\alpha-1) \ldots (\alpha-n+1)$.

Then, $\qquad (1+x)^\alpha = 1 + \alpha x + \frac{\alpha(\alpha-1)}{2!} x^2 + \ldots$

$$\ldots + \frac{\alpha(\alpha-1) \ldots (\alpha-n+2)}{(n-1)!} x^{n-1} + R_n(x),$$

where $\quad R_n(x) = \frac{x^n}{n!} F^{(n)}(\xi).$

Suppose now that $0 \le x < 1$. So

$$|R_n(x)| = \left| \begin{pmatrix} \alpha \\ n \end{pmatrix} x^n (1+\xi)^{\alpha-n} \right| \le K \left| \begin{pmatrix} \alpha \\ n \end{pmatrix} x^n \right|, \qquad \ldots (*)$$

where $K = 2^{\alpha}$ if $\alpha \geq 0$ and $K = 1$ if $\alpha < 0$, since $(1+\xi) \geq 1$. To see that $|R_n(x)| \to 0$ as $n \to \infty$, notice

that $\dfrac{\left| \binom{\alpha}{n+1} x^{n+1} \right|}{\left| \binom{\alpha}{n} x^n \right|} = \dfrac{|n-\alpha|}{n+1} x \to x$ as $n \to \infty$.

Since $0 \leq x < 1$, it follows exactly as in §65 that

$\left| \binom{\alpha}{n} x^n \right| \to 0$ as $n \to \infty$. So it follows from (*) that $R_n(x) \to 0$ as $n \to \infty$. This proves the result for $x \varepsilon [0,1[$.

To prove it for $x \varepsilon]-1,0[$, a slightly more delicate argument using Cauchy's remainder is needed. (Compare the argument for $x \varepsilon]0,1[$ in §174.)

<u>Notice</u>: (i) Unless α is a non-negative integer the <u>series</u> developed in the above result is an <u>infinite</u> series with radius of convergence 1. (See §190.) This means that the series is absolutely convergent for $-1 < x < 1$ and divergent for $|x| > 1$. What happens for $x = \pm 1$ is treated in Ex.12.10.

(ii) The result applies to both rational and irrational powers even although irrational powers are not defined till §210. Differentiation of irrational powers presents no special difficulty: see Ex.10.53.

176. Memorise the following special cases of the general binomial theorem. They are all valid for $-1 < x < 1$.

1. $(1+x)^{-1} = 1 - x + x^2 - x^3 + \ldots$,

2. $(1-x)^{-1} = 1 + x + x^2 + x^3 + \ldots$,

3. $(1-x)^{-2} = 1 + 2x + 3x^2 + 4x^3 + \ldots$,

4. $(1-x)^{-\frac{1}{2}} = 1 + \frac{1}{2}x + \frac{1.3}{2.4}x^2 + \frac{1.3.5}{2.4.6}x^3 + \ldots$.

Notice that (3) follows from (2) on differentiating. The result of §198 justifies this manoeuvre.

177. LEIBNIZ'S THEOREM ON THE NTH DERIVATIVE OF A PRODUCT

While the product rule for differentiation gives a formula for the <u>first</u> derivative of a product of two functions, the following result of Leibniz gives a formula for the nth derivative.

RESULT. Suppose that f and g are n times differentiable at c. Then so also is fg and

$$(fg)^{(n)}(c) = f^{(n)}(c)g(c) + \binom{n}{1}f^{(n-1)}(c)g'(c)$$

$$+ \binom{n}{2}f^{(n-2)}(c)g''(c) + \ldots + f(c)g^{(n)}(c).$$

Proof. For $n = 1$, the result is the product rule proved in §121. Proceed by induction. Assume the above result is true for n and take the derivative of each side using the product rule. So

$$(fg)^{(n+1)}(c) = f^{(n+1)}(c)g(c) + [\binom{n}{1} + \binom{n}{0}]f^{(n)}(c)g'(c)$$

$$+ [\binom{n}{2} + \binom{n}{1}]f^{(n-1)}(c)g''(c)$$

$$+ \ldots + [\binom{n}{n} + \binom{n}{n-1}]f'(c)g^{(n)}(c)$$

$$+ f(c)g^{(n+1)}(c)$$

$$= f^{(n+1)}(c)g(c) + \binom{n+1}{1}f^{(n)}(c)g'(c)$$

$$+ \binom{n+1}{2}f^{(n-1)}(c)g''(c) + \ldots + f(c)g^{(n+1)}(c),$$

using the fact that $\binom{n}{r+1} + \binom{n}{r} = \binom{n+1}{r+1}$.

So the result is then true for $(n+1)$ and so it is true for all $n \in N$ by induction.

178. USE OF LEIBNIZ'S THEOREM IN DEVELOPING A MACLAURIN SERIES

Finding a Maclaurin expansion as in §173 depends on being able to find $f^{(n)}(0)$ for $n \geq 0$. Where $f^{(n)}(x)$ can be found explicitly, as in §174, this is easy. However, in general, $f^{(n)}(x)$ is not easy to calculate explicitly. In this situation Leibniz's theorem can make a useful but limited contribution as the following example illustrates.

EXAMPLE. The function f is defined on $[-1,1]$ by $f(x) = \frac{1}{2}(\sin^{-1}x)^2$. Find its Maclaurin series.

Solution. $f'(x) = \sin^{-1}x/\sqrt{(1-x^2)}$ and so $f'(0) = 0$.

Also $f''(x) = \dfrac{1}{(1-x^2)} + \dfrac{x \sin^{-1}x}{(1-x^2)^{3/2}}$ so that $f''(0) = 1$.

Calculating further derivatives by hand is difficult. Instead aim to find a differential equation satisfied by f but which does not involve $\sin^{-1}x$ explicitly. To this end we can eliminate $\sin^{-1}x$ from the equations for $f'(x)$ and $f''(x)$ to get the differential equation

$$(1-x^2)f''(x) - xf'(x) = 1.$$

Differentiate this n times with Leibniz's theorem to get (for every $n\varepsilon N$)

$$(1-x^2)f^{(n+2)}(x) + \binom{n}{1}(-2x)f^{(n+1)}(x) + \binom{n}{2}(-2)f^{(n)}(x)$$

$$- xf^{(n+1)}(x) - \binom{n}{1}f^{(n)}(x) = 0,$$

i.e. $(1-x^2)f^{(n+2)}(x) - (2n+1)xf^{(n+1)}(x) - n^2f^{(n)}(x) = 0$.

Since we want $f^{(n)}(0)$ etc, put $x = 0$ in this last relation and see that, for $n\varepsilon N$,

$$f^{(n+2)}(0) = n^2f^{(n)}(0). \qquad \ldots(*)$$

Repeated application of $(*)$ gives

$$f^{(3)}(0) = 1^2.f'(0) = 0 \quad \text{from above.}$$

Then $\qquad f^{(5)}(0) = 3^2.f^{(3)}(0) = 0,$

and clearly $\quad f^{(2n+1)}(0) = 0 \quad$ for all $\ n \geq 1$.

On the other hand $f^{(4)}(0) = 2^2.f''(0) = 2^2 \quad$ from above.

Then $\qquad f^{(6)}(0) = 4^2.f^{(4)}(0) = 4^2.2^2,$

$\qquad\qquad f^{(8)}(0) = 6^2.f^{(6)}(0) = 6^2.4^2.2^2,$

and in general, $\quad f^{(2n)}(0) = 2^2.4^2.6^2\ldots(2n-2)^2 \quad (n\varepsilon N)$.

So the Maclaurin series of f is

$$\frac{1}{2!}x^2 + \frac{2^2}{4!}x^4 + \frac{2^2.4^2}{6!}x^6 + \ldots + \frac{2^2.4^2\ldots(2n-2)^2}{(2n)!}x^{2n} + \ldots \ .$$

Notice: (i) The results of §154 may help in writing down the general term in the above expansion.

(ii) In the method of the above example there is no guarantee that the series developed actually converges, and even if it does there is no guarantee that it converges to $f(x)$. (Such guarantees were provided in the method of §174 by considering the remainder term.)

In the above case the series developed converges and <u>does</u> represent $\frac{1}{2}(\sin^{-1}x)^2$ for $x\epsilon[-1,1]$. However, such a happy situation does not obtain in general: a Maclaurin series need not represent the function from which it is calculated. For example, §224(ii) shows how to construct different functions with the same Maclaurin series.

<u>EXAMPLE</u>. Function f is infinitely differentiable on R, $f(0) = 0$, $f'(0) = 1$ and f satisfies the differential equation $(1+x^2)f''(x) + xf'(x) = 0$ on some neighbourhood of 0. Show that $f^{(n+2)}(0) = -n^2f^{(n)}(0)$ $(n\epsilon N)$. Hence show that the Maclaurin series of f is

$$\sum_{n=0}^{\infty} \frac{(-1)^n (2n)! \, x^{2n+1}}{2^{2n}(2n+1)(n!)^2} \, .$$

EXAMPLES 7

1. The function f is <u>even</u> and is infinitely differentiable on R. By <u>differentiating</u> the relation $f(x) = f(-x)$, prove that the Maclaurin series of f consists of <u>even</u> powers of x only.
    Prove the analogue for an odd function. (Notice how these results are borne out by the series for the sine and cosine functions in §216.)

2. Find explicitly the nth derivative of:
(a) $(1-x)^{-1}$, (b) $(1+x)^{-1}$, (c) $(1-x^2)^{-1}$,
(d) $(3-2x)^{-1}$, (e) $x(1+x)^{-1}$, (f) $\log x$, (g) $x^2 e^{2x}$.

3. Let p be a real polynomial function of degree n. By applying Taylor's theorem to p show that
$$p(x) = p(a) + (x-a)p'(a) + \frac{(x-a)^2}{2!}p''(a) + \ldots + \frac{(x-a)^n}{n!}p^{(n)}(a).$$

4. Express $x^4 - 7x^3 + 6x + 2$ as a polynomial in $(x-1)$. Hence express $(x^4 - 7x^3 + 6x + 2)/(x - 1)^5$ in partial fractions.

5. The non-constant function f is periodic and infinitely differentiable on R. Its Maclaurin series is $a_0 + a_1 x + a_2 x^2 + \ldots$ and this converges to $f(x)$ for all $x\epsilon R$. Prove that all terms of the sequence $\{a_0, a_1, a_2, \ldots\}$ cannot have the same sign. (Cases in

point are the series for sine and cosine in §216.   This
matter is taken further in Ex.9.14.)

6.   Use Taylor's theorem to prove that

$$(1-x)^{-2} = 1 + 2x + 3x^2 + 4x^3 + \ldots \ , \text{ for } -1 < x < 1.$$

7.   It is easy to check using the formula for the sum
of a geometric series that, for $x \varepsilon]0,1[$,

$$(1-x)^{-1} = 2 + 2^2(x-\tfrac{1}{2}) + 2^3(x-\tfrac{1}{2})^2 + 2^4(x-\tfrac{1}{2})^3 + \ldots .$$

Derive this result by applying Taylor's theorem to the
function $1/(1-x)$ with $a = \tfrac{1}{2}$.

8.   By writing the series as $(1+x)^\alpha$ for suitable
x and α,   show that $1 + \dfrac{1}{2}\cdot\dfrac{1}{2} + \dfrac{1.3}{2.4}\left(\dfrac{1}{2}\right)^2 + \dfrac{1.3.5}{2.4.6}\left(\dfrac{1}{2}\right)^3 + \ldots = \sqrt{2}.$

9.   By introducing factors above and below show that
$$\frac{1.3.5\ldots(2n-1)}{2.4.6\ldots(2n)} = \binom{2n}{n} 4^{-n} .$$

10.   Use partial fractions to show that the coeffic-
ient of x^n in the expansion of $\dfrac{4x^2-3}{(1-x)(1-2x)^2}$ is
$1 - (n+1)2^{n+2}$.   For what values of x is the expansion
valid?

11.   (Experimental errors) A quantity c is
calculated from two measured quantities a and b by
the formula $c = a/b^2$.   The measurements of a and b
are subject to errors of ±5% and ±10% respectively.   By
considering $a(1 \pm .05)b^{-2}(1 \pm .10)^{-2}$ show that the
calculated value of c could be at most 22% below and
at most 30% above the true value.

12.   An infinitely differentiable function f
satisfies the conditions that $f(0) = 1$, $f'(0) = 0$ and
$f''(x) + 4f(x) = 0$ for all $x \varepsilon R$.   Find the Maclaurin
series of f.

13.   The function f is infinitely differentiable
on $]-1,1[$, $f(0) = 1$, $f'(0) = 1$ and for all $x \varepsilon]-1,1[$
$$(1-x^2)f''(x) - xf'(x) - f(x) = 0.$$
Prove that $f^{(n+2)}(0) = (n^2+1)f^{(n)}(0)$ for all $n \varepsilon N$.
Deduce that the Maclaurin series of f is

$$1 + x + \frac{x^2}{2!} + \frac{(1+1^2)}{3!}x^3 + \frac{(1+2^2)}{3!}x^4 + \frac{(1+1^2)(1+3^2)}{5!}x^5 + \ldots .$$

14.   Write $(1+x+x^2)^{-1}$ as $(1-x)/(1-x^3)$ and expand the denominator to show that, for all $x \varepsilon\,]-1,1[$,

$$(1+x+x^2)^{-1} = \sum_{n=0}^{\infty} a_n x^n , \text{ where } a_n = 1, -1, 0 \text{ according}$$

as $n \equiv 0, 1, 2 \pmod 3$.

15.   By writing each of the following as $(1+x)^{\alpha}$ for suitable values of x and α, find their sums:

(a) $1 + \dfrac{3}{8} + \dfrac{3.9}{8.16} + \dfrac{3.9.15}{8.16.24} + \cdots$,

(b) $1 + \dfrac{2}{9} + \dfrac{2.5}{9.18} + \dfrac{2.5.8}{9.18.27} + \cdots$.

16.   Investigate the series for the function $(1-x)^{-1}$ that Taylor's theorem develops at a general point a, where $a \,\varepsilon\, R-\{1\}$.   Check on the range of values of x for which the series does represent the function by summing geometric series.   (Compare Exs.6 & 7 above.)

VIII.   MORE ON SERIES

---

179.   ESTIMATING THE PARTIAL SUMS OF CERTAIN DIVERGENT SERIES

There are two ways of doing this, one using differentiation, the other using integration, but both achieving the same result.   The following example illustrates the use of integration.   How to use differentiation is indicated in (ii) below.

<u>EXAMPLE</u>.   By considering $\int_1^n 1/x \, dx$, prove that, for all integers $n \geq 2$,

$$\frac{1}{n} + \log n < 1 + \frac{1}{2} + \frac{1}{3} + \dots + \frac{1}{n} < 1 + \log n. \quad \dots (*)$$

Prove also that $1 + \frac{1}{2} + \frac{1}{3} + \dots + \frac{1}{n} - \log n$ decreases as n increases and deduce that it tends to a limit as $n \to \infty$.

<u>Solution</u>.   From §245, $\int_1^n 1/x \, dx$ can be interpreted as the area under the curve $y = 1/x$ between $x = 1$ and

$x = n$.   This area lies between the sums of the areas of the lower and upper rectangles shown. We can calculate the areas of these rectangles by taking base times height and conclude that, for $n \geq 2$,

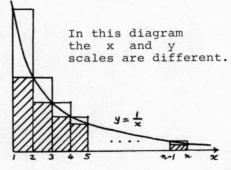

In this diagram the   x   and   y   scales are different.

$y = \frac{1}{x}$

$$1 \cdot \frac{1}{2} + 1 \cdot \frac{1}{3} + \dots + 1 \cdot \frac{1}{n} < \int_1^n 1/x \, dx < 1 \cdot 1 + 1 \cdot \frac{1}{2} + \dots + 1 \cdot \frac{1}{n-1},$$

i.e. $$\frac{1}{2} + \frac{1}{3} + \dots + \frac{1}{n} < \log n < 1 + \frac{1}{2} + \frac{1}{3} + \dots + \frac{1}{n-1}.$$

Adding   1   to the left hand inequality and adding $1/n$ to the right hand inequality gives   (*) as required.

Now let $G(n) = 1 + \frac{1}{2} + \frac{1}{3} + \dots + \frac{1}{n} - \log n$.   Then

$$G(n+1) - G(n) = \frac{1}{n+1} - (\log(n+1) - \log n) < 0, \text{ by the}$$

mean-value theorem as in §134(ii).

So {G(n)} is a decreasing sequence. Also from (*) it is clear that {G(n)} is bounded below by 0. So {G(n)} tends to a limit γ as n → ∞.

Notice: (i) It is actually clear from (*) that γ lies between 0 and 1. In fact γ is called <u>Euler's constant</u> and has the value 0.577... .

(ii) To get the above result using differentiation we start from the function log x and apply the mean-value theorem to it on the intervals [k,k+1] for each kεN. This gives

$$\log(k+1) - \log k = \frac{1}{\xi}, \text{ where } \xi\varepsilon]k,k+1[.$$

It follows that

$$\frac{1}{k+1} < \log(k+1) - \log k < \frac{1}{k} \quad \text{for all } k\varepsilon N.$$

Add up these relations for k = 1, 2, ... , (n-1). So

$$\sum_{k=1}^{n-1} \frac{1}{k+1} < \sum_{k=1}^{n-1} (\log(k+1) - \log k) < \sum_{k=1}^{n-1} \frac{1}{k}.$$

The middle term then collapses to give

$$\frac{1}{2} + \frac{1}{3} + \ldots + \frac{1}{n} < \log n < 1 + \frac{1}{2} + \ldots + \frac{1}{n-1},$$

which is just a stage in the example above and things can proceed as before.

EXAMPLE. Show that $1 + \frac{1}{\sqrt{2}} + \frac{1}{\sqrt{3}} + \ldots + \frac{1}{\sqrt{n}}$ lies between $2\sqrt{n} - 2 + 1/\sqrt{n}$ and $2\sqrt{n} - 1$.

180. THE PARTIAL SUMS OF THE HARMONIC SERIES

Finding the partial sums of a geometric series explicitly (§144) is relatively easy. To do the same for the harmonic series (§146) is much more difficult. Notice however that §179 provides the following estimate.

RESULT. For nεN,

$$1 + \frac{1}{2} + \frac{1}{3} + \ldots + \frac{1}{n} = \log n + \gamma + \alpha_n,$$

where γ is Euler's constant and $\alpha_n \to 0$ as n → ∞.

Proof. See §179.

EXAMPLE. Show that $\lim\limits_{n\to\infty}\left(\dfrac{1}{n+1} + \dfrac{1}{n+2} + \ldots + \dfrac{1}{2n}\right) = \log 2$.

Solution. Let u_n denote the sum $1 + \dfrac{1}{2} + \dfrac{1}{3} + \ldots + \dfrac{1}{n}$.

Then

$$\dfrac{1}{n+1} + \dfrac{1}{n+2} + \ldots + \dfrac{1}{2n} = u_{2n} - u_n$$

$$= (\log 2n + \gamma + \alpha_{2n}) - (\log n + \gamma + \alpha_n)$$

$$= \log 2 + \alpha_{2n} - \alpha_n$$

$$\to \log 2 \quad \text{as} \quad n \to \infty.$$

(This settles the value of L in the example in §57. See also §246.)

181. THE SUM OF THE LOGARITHMIC SERIES

EXAMPLE. The series $1 - \dfrac{1}{2} + \dfrac{1}{3} - \dfrac{1}{4} + \dfrac{1}{5} - \ldots$ is called the <u>logarithmic series</u>. Show that its sum is $\log 2$.

Solution. Let s_n denote the nth partial sum of the given series. Then,

$$s_{2n} = 1 - \dfrac{1}{2} + \dfrac{1}{3} - \dfrac{1}{4} + \ldots + \dfrac{1}{2n-1} - \dfrac{1}{2n}$$

$$= \left(1 + \dfrac{1}{2} + \dfrac{1}{3} + \dfrac{1}{4} + \ldots + \dfrac{1}{2n-1} + \dfrac{1}{2n}\right) - 2\left(\dfrac{1}{2} + \dfrac{1}{4} + \ldots + \dfrac{1}{2n}\right)$$

$$= u_{2n} - u_n \quad \text{(in the notation of §180)}$$

$$\to \log 2 \quad \text{as} \quad n \to \infty.$$

Also $s_{2n+1} - s_{2n} \to 0$ as $n \to \infty$. So $s_n \to \log 2$ as $n \to \infty$. Hence the result.

182. THE INTEGRAL TEST FOR CONVERGENCE OF SERIES

In some cases the convergence/divergence question for series of positive terms can be related to the convergence/divergence question for integrals. The following result does this.

RESULT. (Integral test) Let f be a positive decreasing integrable function on $[1,\infty[$. Let $F(n) = \int_1^n f(x)\,dx$. Then $\sum f(n)$ converges or diverges according as $\{F(n)\}$ tends to a finite limit or tends to infinity as $n \to \infty$, i.e. according as $\int_1^\infty f(x)\,dx$ converges or diverges.

Proof. From area considerations notice first that $\{F(n)\}$ increases with n since f is positive. So by §57 conclude that $\{F(n)\}$ either tends to a finite limit or to infinity as $n \to \infty$. Also, from area considerations as in §179, and by using the fact that f is decreasing, notice that

$$f(2) + f(3) + \ldots + f(n) \leq F(n) \leq f(1) + f(2) + \ldots + f(n-1).$$

If $\{F(n)\}$ tends to a finite limit as $n \to \infty$, use the left hand inequality to conclude that $\sum f(n)$ converges. If $F(n) \to \infty$ as $n \to \infty$, use the right hand inequality to conclude that $\sum f(n)$ diverges.

EXAMPLE. Use the integral test to show that both

$$\sum \frac{1}{n \log n} \quad \text{and} \quad \sum \frac{1}{n \log n \, \log\log n} \quad \text{diverge}.$$

OPERATIONS ON SERIES

183. §142 warned against doing certain operations on series in general. However, within the set of all <u>absolutely convergent series</u> we can (a) rearrange without change of sum (§184), and (b) multiply two series together, the product series converging to the product of the sums of the component series (§186).

184. REARRANGEMENTS OF A SERIES

$\sum b_n$ only qualifies as a <u>rearrangement</u> of $\sum a_n$ if each term in $\sum a_n$ is assigned a definite place in the rearrangement. This means that $b_n = a_{f(n)}$ for each $n \in N$, where $f : N \to N$ is a bijective function. So no term can be deferred indefinitely in the rearrangement.

RESULT. Every rearrangement of an absolutely convergent series is convergent to the same sum.

Proof. Let $\sum a_n$ and $\sum a_{f(n)}$ be the original and rearranged series with respective partial sums s_n and t_n.

Choose $\varepsilon > 0$. Now the partial sums of $\sum |a_n|$ increase to a limit as $n \to \infty$, because $\sum a_n$ is absolutely convergent. So we can choose integer N such that

$$\sum_{k=N+1}^{\infty} |a_k| < \varepsilon. \qquad \ldots(*)$$

Now choose integer P such that t_p includes all the terms $a_1, a_2, \ldots, a_N$. Clearly $P \geq N$. Then, for every integer $n \geq P$, we know that the terms $a_1, a_2, \ldots, a_N$ occur both in t_n and in s_n, so that they cancel out in $(t_n - s_n)$. So, for every $n \geq P$,

$$|t_n - s_n| = \left| \sum_{i \varepsilon C} a_i - \sum_{j \varepsilon D} a_j \right|,$$

where C, D are finite sets of integers greater than N,

$$\leq \sum_{i \varepsilon C} |a_i| + \sum_{j \varepsilon D} |a_j|$$

$$\leq \sum_{i=N+1}^{\infty} |a_i| + \sum_{j=N+1}^{\infty} |a_j|$$

$$< 2\varepsilon \qquad \text{from } (*).$$

So $(t_n - s_n) \to 0$ as $n \to \infty$, and hence since $t_n = (t_n - s_n) + s_n$, it follows that t_n tends to the same limit as s_n as $n \to \infty$, i.e. the series have the same sum.

185. EXAMPLES OF REARRANGEMENT OF SERIES

On one hand notice that the result of §184 tells us that every rearrangement of the <u>absolutely convergent</u> series $1 - \frac{1}{2^2} + \frac{1}{3^2} - \frac{1}{4^2} + \frac{1}{5^2} - \frac{1}{6^2} + \ldots$, which incidentally has the sum $\pi^2/12$, also has the sum $\pi^2/12$.
On the other hand notice that a <u>conditionally convergent</u> series like $1 - \frac{1}{2} + \frac{1}{3} - \frac{1}{4} + \frac{1}{5} - \frac{1}{6} + \ldots$,

which has sum $\log 2$, is not protected (with regard to rearrangement) by the result of §184 and it can in fact be rearranged to give other sums or to diverge as the following example shows.

EXAMPLE.   (Rearrangements of the logarithmic series)
Prove that

(a) $1 + \frac{1}{3} + \frac{1}{5} - \frac{1}{2} + \frac{1}{7} + \frac{1}{9} + \frac{1}{11} - \frac{1}{4} + \ldots$ converges to $\log 2\sqrt{3}$,

(b) $1 - \frac{1}{2} + \frac{1}{3} + \frac{1}{5} - \frac{1}{4} + \frac{1}{7} + \frac{1}{9} + \frac{1}{11} - \frac{1}{6} + \ldots$ diverges.

(In (a) the pattern of signs is 3+, 1-, 3+, 1-, etc.
In (b) it is 1+, 1-, 2+, 1-, 3+, 1-, 4+, 1-, etc.)

Solution.   (a)   Let t_n denote the nth partial sum
of the rearrangement.   Consider the first n batches
of terms, i.e. the first 4n terms.   So

$$t_{4n} = (1 + \tfrac{1}{3} + \tfrac{1}{5} - \tfrac{1}{2}) + (\tfrac{1}{7} + \tfrac{1}{9} + \tfrac{1}{11} - \tfrac{1}{4}) +$$

$$\ldots + (\tfrac{1}{6n-5} + \tfrac{1}{6n-3} + \tfrac{1}{6n-1} - \tfrac{1}{2n})$$

$$= (1 + \tfrac{1}{2} + \tfrac{1}{3} + \ldots + \tfrac{1}{6n-1} + \tfrac{1}{6n})$$

$$-(\tfrac{1}{2} + \tfrac{1}{4} + \ldots + \tfrac{1}{6n}) - (\tfrac{1}{2} + \tfrac{1}{4} + \ldots + \tfrac{1}{2n})$$

$$= u_{6n} - \tfrac{1}{2}u_{3n} - \tfrac{1}{2}u_n \quad \text{(in the notation of §180)}$$

$$= (\log 6n + \gamma) - \tfrac{1}{2}(\log 3n + \gamma) - \tfrac{1}{2}(\log n + \gamma) + \beta_n,$$
$$\text{where } \beta_n \to 0 \text{ as } n \to \infty,$$

$$\to \log 2\sqrt{3} \quad \text{as } n \to \infty.$$

Also t_{4n+1}, t_{4n+2}, t_{4n+3} differ from t_{4n} by amounts
which tend to zero as $n \to \infty$.   So $t_n \to \log 2\sqrt{3}$ as
$n \to \infty$ and the result is proved.

(b)   Here the first n batches of terms include
$(1+2+3+\ldots+n)$, i.e. $\frac{1}{2}n(n+1)$, positive terms and
n negative terms.   The last positive term is $1/k$,
where $k = 2(\frac{1}{2}n(n+1)) - 1$, i.e. $k = n^2+n-1$, and the last
negative term is $1/2n$.   So if E_n denotes the sum of
all terms in these n batches,

$$E_n = (1 + \tfrac{1}{3} + \tfrac{1}{5} + \ldots + \tfrac{1}{n^2+n-1}) - (\tfrac{1}{2} + \tfrac{1}{4} + \ldots + \tfrac{1}{2n})$$

$$> (\tfrac{1}{2} + \tfrac{1}{4} + \tfrac{1}{6} + \ldots + \tfrac{1}{n^2+n}) - (\tfrac{1}{2} + \tfrac{1}{4} + \ldots + \tfrac{1}{2n})$$

$$= \tfrac{1}{2}(u_{\frac{1}{2}n(n+1)} - u_n), \quad \text{and this behaves like}$$

$\frac{1}{2}(\log \frac{1}{2}n(n+1) - \log n)$ as $n \to \infty$, by §180,
i.e. like $\frac{1}{2}\log \frac{1}{2}(n+1)$.   So $E_n \to \infty$ and (b) diverges.

Notice: (i) In each case we looked at n complete batches of terms.

(ii) Checking the behaviour of t_{4n+1}, t_{4n+2}, t_{4n+3} is needed in (a) to conclude convergence: the series $1 - 1 + 1 - 1 + 1 - 1 + \ldots$ illustrates that if t_{4n} tends to a limit as $n \to \infty$ this does not guarantee that the series converges.

186. MULTIPLICATION OF SERIES

Notice that

$$(a_0 + a_1 x + a_2 x^2 + \ldots)(b_0 + b_1 x + b_2 x^2 + \ldots)$$
$$= a_0 b_0 + (a_1 b_0 + a_0 b_1)x + (a_2 b_0 + a_1 b_1 + a_0 b_2)x^2$$
$$+ (a_3 b_0 + a_2 b_1 + a_1 b_2 + a_0 b_3)x^3 + \ldots .$$

This motivates the form of the general term in the Cauchy product of two series.

Given two series $\sum a_n$ $(n \geq 0)$ and $\sum b_n$ $(n \geq 0)$, take, for $n \geq 0$, $c_n = a_n b_0 + a_{n-1} b_1 + a_{n-2} b_2 + \ldots + a_0 b_n$. Then the series $\sum c_n$ is called the Cauchy product of the two original series.

In the figure successive terms of the product series consist of sums along the diagonals shown. The following result is vital to our development of the exponential function later.

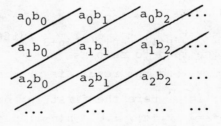

RESULT. Let $\sum a_n$ and $\sum b_n$ be absolutely convergent series with sums A and B. Then the Cauchy product series $\sum c_n$ converges to AB.

Proof. Let A_n, A_n', B_n, B_n', C_n, D_n denote the nth partial sums of the series with nth terms a_n, $|a_n|$, b_n, $|b_n|$, c_n (as defined above) and d_n where $d_n = |a_n||b_0| + |a_{n-1}||b_1| + \ldots + |a_0||b_n|$. Let A', B' denote the sums of $\sum |a_n|$ and $\sum |b_n|$.

Since $A_n' \to A'$ and $B_n' \to B'$, it follows that $A_n'B_n' \to A'B'$ as $n \to \infty$.

Choose any integer $n \geq 2$. Use [ ] to denote the integral part function. Then if the sets of all terms $|a_i||b_k|$ occurring in $A_{[\frac{1}{2}n]}'B_{[\frac{1}{2}n]}'$, D_n and $A_n'B_n'$ are respectively X, Y and Z, it is fairly clear that $X \subseteq Y \subseteq Z$. But then since all terms $|a_i||b_k|$ are non-negative, it follows that

$$A_{[\frac{1}{2}n]}'B_{[\frac{1}{2}n]}' \leq D_n \leq A_n'B_n' .$$

In this relation both left and right hand sides tend to $A'B'$ as $n \to \infty$, so that $D_n \to A'B'$ as $n \to \infty$ and more usefully for us $(D_n - A_n'B_n') \to 0$ as $n \to \infty$.

Now notice that the suffices (i,k) of terms a_ib_k that occur in $(C_n - A_nB_n)$ are the same as those of the terms $|a_i||b_k|$ that occur in $(D_n - A_n'B_n')$. So by the triangle inequality it follows that

$$|C_n - A_nB_n| \leq |D_n - A_n'B_n'|$$
$$\to 0 \quad \text{as} \quad n \to \infty.$$

So, $C_n = (C_n - A_nB_n) + A_nB_n \to 0 + AB = AB$ as $n \to \infty$. This proves the result.

Other results guarantee the convergence of the product series with less stringent demands on the components, (e.g. a result of Mertens), but notice that in general the product of two conditionally convergent series need not be convergent. See Ex.8.15.

187. AN EXAMPLE ON MULTIPLICATION OF SERIES

<u>EXAMPLE</u>. Show that, for $-1 < x < 1$,

$$\left(\sum_{n=1}^{\infty} nx^n \right)^2 = \frac{1}{6} \sum_{n=2}^{\infty} n(n^2 - 1)x^n .$$

<u>Solution</u>. It is easy to check that $\sum nx^n$ is absolutely convergent for $-1 < x < 1$, using the ratio test or otherwise. So squaring the series is justified for these values of x by §186. Now

$$\left(\sum_{n=1}^{\infty} nx^n \right)^2 = \left(\sum_{r=1}^{\infty} rx^r \right) \left(\sum_{s=1}^{\infty} sx^s \right) = \sum_{r=1}^{\infty} \sum_{s=1}^{\infty} rs\, x^{r+s} .$$

Collect all terms of degree n in the product. This gives the general term c_n $(n \geq 2)$, where

$$c_n = \sum_{r=1}^{n-1} rs\, x^{r+s} \quad , \text{ where } r+s = n, \text{ i.e. } s = n-r,$$

$$= x^n \sum_{r=1}^{n-1} r(n-r) = x^n \left(n \sum_{r=1}^{n-1} r - \sum_{r=1}^{n-1} r^2 \right)$$

$$= x^n \left[\tfrac{1}{2} n^2 (n-1) - \tfrac{1}{6} n(n-1)(2n-1) \right] \quad \text{(by Ex.1.14)}$$

$$= \tfrac{1}{6} n(n^2 - 1) x^n, \quad \text{as required.}$$

Notice that it is a good idea to start with <u>different</u> letters r and s as suffices in the <u>component</u> series. Then pursue the general term in the product and this generally forces a relation between r and s as in the example above.

<u>EXAMPLE</u>. Prove that, for $-1 < x < 1$,

$$\left(\sum_{n=1}^{\infty} nx^n \right) \left(\sum_{n=1}^{\infty} n^2 x^n \right) = \sum_{n=2}^{\infty} \frac{1}{12} n^2 (n^2 - 1) x^n.$$

See also §201 and Ex.10.38 for other examples of multiplication of absolutely convergent series.

EXAMPLES 8

1. By applying the mean-value theorem to $\log x$ on $[2k-1, 2k+1]$, prove that for all positive integers k

$$\frac{2}{2k+1} < \log(2k+1) - \log(2k-1) < \frac{2}{2k-1}.$$

Deduce that $1 + \frac{1}{3} + \frac{1}{5} + \ldots + \frac{1}{2n-1}$ lies between $\tfrac{1}{2}\log(2n-1)$ and $1 + \tfrac{1}{2}\log(2n-1)$.

2. Show that

$$\frac{1}{2 \log 2} + \frac{1}{3 \log 3} + \ldots + \frac{1}{n \log n} - \log\log n$$

tends to a limit as $n \to \infty$.

3. Let $f(x) = x \log x$ $(x > 0)$. Prove using the mean-value theorem that, for all $n \in \mathbb{N}$,

$$1 + \log n < f(n+1) - f(n) < 1 + \log(n+1).$$

Deduce, by a collapsing argument, that, for all $n \geq 2$,

$$n + \log(n-1)! < 1 + n \log n < n + \log n!.$$

Deduce that n! lies between $e(n/e)^n$ and $ne(n/e)^n$. (This last result is a poor version of Stirling's formula for n! in §270.) Hence prove that

$(n!)^{1/n}/n \to 1/e$ as $n \to \infty$.

4. Use the integral test to discuss the convergence of $\sum \dfrac{1}{n(\log n)^\alpha}$ for all values of $\alpha > 0$.

5. Discuss the convergence of $\sum \dfrac{1}{n \log n \, (\log\log n)^\alpha}$ for all $\alpha > 0$.

6. Find the limits as $n \to \infty$ of

(a) $\dfrac{1}{n+1} + \dfrac{1}{n+2} + \cdots + \dfrac{1}{3n}$, (b) $\dfrac{1}{2n+1} + \dfrac{1}{2n+2} + \cdots + \dfrac{1}{3n}$,

(c) $\dfrac{1}{n^2+1} + \dfrac{1}{n^2+2} + \cdots + \dfrac{1}{n^2+2n+1}$

7. By doing estimation and using sandwich results, find the limits as $n \to \infty$ of

(a) $\dfrac{1}{2n+1} + \dfrac{1}{2n+3} + \cdots + \dfrac{1}{4n-1}$, (b) $\dfrac{1}{4n+1} + \dfrac{1}{4n+5} + \cdots + \dfrac{1}{8n-3}$,

(c) $\dfrac{1}{\sqrt{(n(n+1))}} + \dfrac{1}{\sqrt{((n+1)(n+2))}} + \cdots + \dfrac{1}{\sqrt{((2n-1)(2n))}}$,

(d) $\dfrac{n}{(n+1)^2} + \dfrac{n+1}{(n+2)^2} + \cdots + \dfrac{2n-1}{(2n)^2}$.

8. (Rearrangements of the logarithmic series) Prove that

(a) $1 + \dfrac{1}{3} - \dfrac{1}{2} + \dfrac{1}{5} + \dfrac{1}{7} - \dfrac{1}{4} + \dfrac{1}{9} + \dfrac{1}{11} - \dfrac{1}{6} + \cdots = \dfrac{3}{2} \log 2$,

(b) $1 - \dfrac{1}{2} - \dfrac{1}{4} + \dfrac{1}{3} - \dfrac{1}{6} - \dfrac{1}{8} + \dfrac{1}{5} - \dfrac{1}{10} - \dfrac{1}{12} + \cdots = \dfrac{1}{2} \log 2$,

(c) $1 - \dfrac{1}{2} - \dfrac{1}{4} - \dfrac{1}{6} - \dfrac{1}{8} + \dfrac{1}{3} - \dfrac{1}{10} - \dfrac{1}{12} - \dfrac{1}{14} - \dfrac{1}{16} + \cdots = 0$.

9. The conditionally convergent series

$$1 - 1 + \dfrac{1}{2} - \dfrac{1}{2} + \dfrac{1}{3} - \dfrac{1}{3} + \cdots$$

is rearranged by taking the first p positive terms, then the first q negative terms, then the next p positive terms, then the next q negative terms and so on. Prove that this rearrangement converges to $\log(p/q)$. [Hint: Consider n batches of terms.]

10. Ex.7.6 derives from Taylor's theorem the result that, for $|x| < 1$,

$$(1-x)^{-2} = 1 + 2x + 3x^2 + 4x^3 + \dots .$$

Prove this result by squaring the series for $(1-x)^{-1}$.

11. It is given that, for $x \in [-1,1[$,

$$- \log(1 - x) = \sum_{n=1}^{\infty} \frac{x^n}{n} .$$

Show that this series is absolutely convergent for $|x| < 1$ and deduce that, for $|x| < 1$,

$$(\log(1 - x))^2 = \sum_{n=2}^{\infty} \frac{2}{n} \left(1 + \frac{1}{2} + \frac{1}{3} + \dots + \frac{1}{n-1} \right) x^n .$$

12. Suppose that $\sum a_n z^n$ converges absolutely for $|z| < 1$. Let $s_n = a_0 + a_1 + a_2 + \dots + a_n$. Show that for $|z| < 1$,

$$\left(\sum_{n=0}^{\infty} z^n \right) \left(\sum_{n=0}^{\infty} a_n z^n \right) = \sum_{n=0}^{\infty} s_n z^n .$$

13. Prove that the series $1 - 1 + \frac{1}{2} + \frac{1}{3} - \frac{1}{2} - \frac{1}{3} + \frac{1}{4} + \frac{1}{5} + \frac{1}{6} - \frac{1}{4} - \frac{1}{5} - \frac{1}{6} + \dots$ converges to 0. (Here the nth batch of terms consists of n positive terms followed by n negative terms.)

14. Prove that the series $1 - 1 + \frac{1}{2} + \frac{1}{3} - \frac{1}{2} - \frac{1}{3} + \frac{1}{4} + \frac{1}{5} + \frac{1}{6} + \frac{1}{7} - \frac{1}{4} - \frac{1}{5} - \frac{1}{6} - \frac{1}{7} + \dots$ is <u>not</u> convergent. (Here the nth batch of terms consists of 2^{n-1} positive terms followed by 2^{n-1} negative terms.)

15. (<u>The Cauchy product of two conditionally convergent series need not be convergent.</u>) Denote the square of the series $1 - \frac{1}{\sqrt{2}} + \frac{1}{\sqrt{3}} - \frac{1}{\sqrt{4}} + \frac{1}{\sqrt{5}} - \dots$ (using the Cauchy product) by $c_1 + c_2 + c_3 + c_4 + c_5 + \dots$. Prove that

$$c_{2n} = - \sum_{r=1}^{2n} \frac{1}{\sqrt{(r(2n+1-r))}} .$$

Use the result of Ex.1.31, to show that $c_{2n} \nrightarrow 0$ as $n \to \infty$. Deduce that the product of two conditionally convergent series can diverge.

IX. PROPERTIES OF POWER SERIES

---

188. Taylor's theorem gives a method by which a <u>known</u> function can be expanded as a power series: <u>for</u> example, in §174 the function $1/(1-x)$ is expanded as $1 + x + x^2 + x^3 + \ldots$. There is however the possibility of starting with a power series (i.e. a series of powers of $x \varepsilon R$ or of $z \varepsilon C$) and studying the function defined by the sum of the series, wherever it converges. This proves a powerful method for extending the types of function at our disposal. In this way the exponential, trigonometric and hyperbolic functions all come within range. With them they bring their inverses — the logarithmic, inverse trigonometric and inverse hyperbolic functions.

To be able to handle these particular cases in Chapter 10, we check now on the convergence of power series and also on the continuity and differentiability of the functions they define. Treating only <u>real</u> series here gives a distorted picture. So we deal first with the full generality of the complex case and specialise to R later.

189. POWER SERIES

These are series of the form $\sum a_n z^n$, where $z \varepsilon C$ and the coefficients a_n $(n \geq 0)$ are fixed complex numbers. The sum of the series varies with z and thereby defines a function wherever it converges. Here it is convenient to start the series with $n = 0$ rather than with $n = 1$ as in Chapter 6. The results of Chapter 6 are easily adapted to the new situation.

190. CIRCLE OF CONVERGENCE

The power series $1 + z + z^2 + z^3 + \ldots$ (§167) converges only inside the circle $|z| = 1$. Such a <u>circle</u> (<u>not</u> a square or any other shape) exists for <u>every</u> power series as the rest of this section shows.

<u>RESULT</u>. Suppose that $\sum a_n z^n$ converges for $z = b$, where $b \neq 0$. Then the series is absolutely convergent on the whole of the open disc $\{w: |w| < |b|\}$.

<u>Proof</u>. Since $\sum a_n b^n$ converges, $a_n b^n \to 0$ as $n \to \infty$. So there exists N such that $|a_n b^n| \leq 1$ for all $n \geq N$.

For all $w \varepsilon C$, $|a_n w^n| = |a_n b^n| \cdot \left|\dfrac{w}{b}\right|^n \leq \left|\dfrac{w}{b}\right|^n$ for all $n \geq N$.

But, for $|w| < |b|$, $\sum \left|\dfrac{w}{b}\right|^n$ is a convergent geometric series, so that $\sum |a_n w^n|$ converges by comparison. So $\sum a_n z^n$ converges absolutely on the disc stated.

Notice: (i)   Each non-zero point $b \varepsilon C$ at which $\sum a_n z^n$
converges provides an open disc $\{z: |z| < |b|\}$ on which
$\sum a_n z^n$ is absolutely convergent.   All such discs have
centre at   0   so that their union is clearly an open disc
with centre at   0   and on which $\sum a_n z^n$ is absolutely
convergent.   Call this disc   D.   In general the radius
of   D   is a positive number   R   but in extreme cases
D   can   (a) degenerate to the origin only (R = 0), or
(b) expand to the whole of the complex plane (R infinite).

The circle $\{z: |z| = R\}$ is called the <u>circle of
convergence</u> and   R   is called the <u>radius of convergence</u>
of the series.

(ii)   For $|z| < R$, $\sum a_n z^n$ is clearly absolutely
convergent.

For $|z| > R$, $\sum a_n z^n$ must be divergent.   If not
and there were a point   b   with $|b| > R$ and $\sum a_n b^n$
convergent, then the above result would give a circle of
convergence with radius greater than   R.

When $|z| = R$, $\sum a_n z^n$ may converge or diverge.
For illustrations of this see §192.

191.   FINDING THE RADIUS OF CONVERGENCE

The ratio test often provides a way of doing this.

<u>RESULT</u>.   Suppose that $|a_{n+1}/a_n| \to L$ $(L \neq 0)$ as $n \to \infty$.
Then $\sum a_n z^n$ has radius of convergence $1/L$.

<u>Proof</u>.   Clearly $\dfrac{|a_{n+1} z^{n+1}|}{|a_n z^n|} \to L|z|$ as $n \to \infty$.   So, by

the ratio test, $\sum a_n z^n$ is absolutely convergent provided
that $L|z| < 1$,   i.e. $|z| < 1/L$.

On the other hand, when $|z| > 1/L$, $L|z| > 1$ and
$\sum |a_n z^n|$ diverges by the ratio test.   So, for $|z| > 1/L$,
$\sum a_n z^n$ is <u>not</u> absolutely convergent and such   z   cannot
lie inside the circle of convergence, by (ii) of §190.

Hence deduce that the radius of convergence is $1/L$.

Notice that similar arguments show that   R   is   0
when $|a_{n+1}/a_n|$ tends to infinity, (e.g. $\sum n! z^n$),   and

R is infinite when $|a_{n+1}/a_n|$ tends to zero, (e.g. $\sum z^n/n!$).

EXAMPLE. Find the radius of convergence of $\sum \frac{(n!)^2}{(2n)!} z^n$.

Solution. $|a_{n+1}/a_n| = \frac{(n+1)(n+1)}{(2n+1)(2n+2)} \to \frac{1}{4}$ as $n \to \infty$.
So the radius of convergence is 4. (So we know at once that the series converges for $|z| < 4$ and diverges for $|z| > 4$.)

192. CONVERGENCE OF A POWER SERIES ON ITS CIRCLE OF CONVERGENCE

The following examples show that a power series may converge at all, some or none of the points on its circle of convergence.

(a) $\sum z^n/n^2$ has radius of convergence 1. It converges at all points z with $|z| = 1$ by comparison with $\sum 1/n^2$.

(b) $\sum z^n/n$ has radius of convergence 1. It diverges for $z = 1$ (harmonic series) and converges for $z = -1$ (logarithmic series). In §275 the harder question of what happens for other z with $|z| = 1$ is resolved.

(c) $\sum z^n$ also has radius of convergence 1. However it diverges at all points z with $|z| = 1$ since the nth term does not tend to zero as $n \to \infty$.

EXAMPLE. Show that the series in (b) converges for $z = \pm i$ by the method suggested in Ex.6.16.

TRUNCATION OF POWER SERIES

193. 0 NOTATION

This highly successful notation achieves suppression of all but the dominant term or terms of a large expression. In truncating power series we find it invaluable. 0 stands for order of magnitude.

Definition. Suppose f and g are functions defined on a subset D of C. Suppose $A \subsetneq D$. Write

$$f(z) = O(g(z)) \quad \text{for all} \quad z \varepsilon A \quad \text{if and only if}$$

there exists a constant K such that $|f(z)| \leq K|g(z)|$ for all $z \varepsilon A$.

<u>EXAMPLES</u>. (a) $2z^2 - 6z^3 + z^4 = O(z^2)$ for $|z| < 1$.
For $|2z^2 - 6z^3 + z^4| \leq 2|z|^2 + 6|z|^3 + |z|^4$
$$\leq 2|z|^2 + 6|z|^2 + |z|^2 \quad \text{for } |z| < 1$$
$$\leq 9|z|^2 \quad \text{for } |z| < 1.$$
(Here $K = 9$ and $A = \{z: |z| < 1\}$.

  (b) $2z^2 - 6z^3 + z^4 = O(z^4)$ for $|z| > 1$.
For $|2z^2 - 6z^3 + z^4| \leq 2|z|^2 + 6|z|^3 + |z|^4$
$$\leq 2|z|^4 + 6|z|^4 + |z|^4 \quad \text{for } |z| > 1$$
$$\leq 9|z|^4 \quad \text{for } |z| > 1.$$
(Here $K = 9$ and $A = \{z: |z| > 1\}$.

Notice from these last two examples that it is important to state the values of z for which the result holds. The constant K must be the same for all such z.

The statement, $f(z) = 1 + 3z + O(z^2)$ for all $z \varepsilon A$, means that $(f(z) - 1 - 3z) = O(z^2)$ for all $z \varepsilon A$. Such use permits algebraic manipulation. For example, suppose that, for all $z \varepsilon A$,

$$f(z) = 1 + 3z + O(z^2) \quad \text{and} \quad g(z) = 2 + 7z + O(z^2).$$

Then $\qquad f(z) + g(z) = 3 + 10z + O(z^2)$

$\qquad$ and $\qquad f(z)g(z) = 2 + 13z + O(z^2)$ for all $z \varepsilon A$.

The notation can clearly be used for real variables too. See §265 and §266.

194. TRUNCATING A POWER SERIES

Often there is a need to approximate a function by the first few terms of its power series. The following result puts a bound on the 'tail' of the series.

<u>RESULT</u>. Let $\sum a_n z^n$ have (possibly infinite) radius of convergence R. Let $0 < r < R$ and let $k \varepsilon N$. Then

$$\sum_{n=0}^{\infty} a_n z^n = \sum_{n=0}^{k-1} a_n z^n + O(z^k) \quad \text{for } |z| \leq r.$$

(Equivalently, there exists $M > 0$ such that

$$\left| \sum_{n=k}^{\infty} a_n z^n \right| \leq M|z|^k \quad \text{for all } z \text{ with } |z| \leq r.)$$

Proof. Notice first that the point $z = r$ lies inside the circle of convergence, so that the series $\sum a_n z^n$ is absolutely convergent at $z = r$. So there exists a number C such that $\sum\limits_{n=k}^{\infty} \left| a_n r^n \right| = C$. Then

$$\left| \sum_{n=k}^{\infty} a_n z^n \right| = \left| a_k z^k + a_{k+1} z^{k+1} + a_{k+2} z^{k+2} + \ldots \right|$$

$$= \left| z^k \right| \left| a_k + a_{k+1} z + a_{k+2} z^2 + \ldots \right|$$

$$\leq \left| z^k \right| \left(\left| a_k \right| + \left| a_{k+1} r \right| + \left| a_{k+2} r^2 \right| + \ldots \right)$$
$$\text{for} \quad |z| \leq r$$

$$\leq \frac{\left| z^k \right|}{r^k} \left(\left| a_k r^k \right| + \left| a_{k+1} r^{k+1} \right| + \left| a_{k+2} r^{k+2} \right| + \ldots \right.$$
$$\text{for} \quad |z| \leq r$$

$$= C \left| z^k \right| / r^k \quad \text{for} \quad |z| \leq r$$

$$= M \left| z^k \right| \quad \text{for} \quad |z| \leq r.$$

Notice that M is the same for all values of z with $|z| \leq r$. This proves the result.

195. USE OF TRUNCATION OF POWER SERIES IN THE EVALUATION OF CERTAIN LIMITS

The following example illustrates a use of §194. It also makes use of the expansions of $\sin x$ and $\cos x$ from §216. These are valid for all $x \in R$.

EXAMPLE. Show that $\lim\limits_{x \to 0} \dfrac{\sin x - x \cos x}{x^3} = \dfrac{1}{3}$.

Solution. Notice that because the denominator has the value 0 at $x = 0$ a quotient result like that of §51 is no help. Using the power series for $\sin$ and $\cos$, we get

$$\frac{\sin x - x \cos x}{x^3} = \frac{(x - x^3/6 + O(x^5)) - x(1 - x^2/2 + O(x^4))}{x^3}$$
$$\text{for} \quad |x| \leq 1, \text{ say,}$$

$$= \frac{1}{x^3} \left[-\frac{x^3}{6} + O(x^5) + \frac{x^3}{2} - O(x^5) \right] \text{for} \quad |x| \leq 1$$

$$= \frac{1}{x^3} \left[\frac{x^3}{3} + O(x^5) \right] \quad \text{for} \quad |x| \leq 1$$

$$= \frac{1}{3} + O(x^2) \to \frac{1}{3} \quad \text{as} \quad x \to 0.$$

Notice that knowing where to truncate the series is a matter of experience. In general truncating too late is better than truncating too early.

EXAMPLE. Show that $\lim\limits_{x \to 0} \dfrac{(1-x)^{-1/3} - (1+x)^{1/3}}{x^2} = \dfrac{1}{3}$.

196. UNIQUENESS OF THE TAYLOR AND MACLAURIN EXPANSIONS

In §174 it was found that the power series $1 + x + x^2 + \ldots$ represents the function $1/(1-x)$ for $-1 < x < 1$. The following result shows that it is impossible for any other expansion in powers of x to represent $1/(1-x)$ on an open interval containing the origin.

RESULT. Suppose that $\sum a_n x^n$ and $\sum b_n x^n$ both represent the function f on an open interval containing the origin. Then $a_n = b_n$ for all $n \geq 0$.

Proof. Suppose that, for all $x \in]-r,r[$,

$$f(x) = \sum_{n=0}^{\infty} a_n x^n = \sum_{n=0}^{\infty} b_n x^n .$$

So $\sum\limits_{n=0}^{\infty} (a_n - b_n) x^n = 0$ for all $x \in]-r,r[$.

Now suppose that a_N and b_N are the first coefficients in the two expansions to differ. So, by §194,

$$0 = (a_N - b_N) x^N + O(x^{N+1}) \quad \text{for all} \quad x \quad \text{in}$$
$$\text{an open interval} \quad I \quad \text{containing} \quad 0.$$

So, $0 = (a_N - b_N) + O(x)$ for all $x \in I-\{0\}$.
Now let $x \to 0$ and deduce that $0 = a_N - b_N$, i.e. $a_N = b_N$.
So the two expansions must be identical.

Obviously a similar result holds for the complex case.

CONTINUITY AND DIFFERENTIABILITY OF POWER SERIES

197. INVARIANCE OF THE RADIUS OF CONVERGENCE UNDER TERM BY TERM DIFFERENTIATION

Term by term differentiation of the series

$$1 + x + x^2 + x^3 + x^4 + \ldots \quad \text{gives} \quad 1 + 2x + 3x^2 + 4x^3 + \ldots \ .$$

Notice that both these series have radius of convergence equal to 1. This is normal: a power series has the same radius of convergence as its term by term derivative.

RESULT. The series $\sum a_n z^n$ and its term by term derivative $\sum n a_n z^{n-1}$ have the same radius of convergence.

<u>Proof</u>.   Consider a particular positive real number r.
Notice that, for every   c   with $0 < c < r$,

$$|a_n c^n| \le |a_n r^n| \le r.|na_n r^{n-1}| \quad \text{for all} \quad n \varepsilon N.$$

So if $\sum |na_n r^{n-1}|$ converges then by comparison so also
does $\sum |a_n c^n|$ for every   c   with $0 < c < r$.   ...(1)

Notice also that, for every   c   with $0 < c < r$, we
can write $r = (1 + \delta)c$ for some $\delta > 0$.   Then since
$n\delta < (1 + \delta)^n$ by §61, deduce that

$$n\delta c^n < ((1 + \delta)c)^n, \quad \text{i.e.} \quad \delta c.nc^{n-1} < ((1 + \delta)c)^n.$$

$$\text{So} \quad |na_n c^{n-1}| < \tfrac{1}{\delta c}|a_n r^n|.$$

So, if $\sum |a_n r^n|$ converges then by comparison so also
does $\sum |na_n c^{n-1}|$ for every   c   with $0 < c < r$.   ...(2)

Together (1) and (2) show that the radii of convergence are equal.

198.   DIFFERENTIABILITY OF THE SUM FUNCTION OF A POWER SERIES

Restrict attention to real power series, i.e.
$\sum a_n x^n$, where $a_n \varepsilon R$ and $x \varepsilon R$.   Notice that, for
$-1 < x < 1$,

$$(1 - x)^{-1} = 1 + x + x^2 + x^3 + \dots .$$

That it is permissible to differentiate both sides of
this relation to obtain the result that, for $-1 < x < 1$,

$$(1 - x)^{-2} = 1 + 2x + 3x^2 + 4x^3 + \dots$$

is proved by the following result.

<u>RESULT</u>.   Suppose that $\sum a_n x^n$ has (possibly infinite)
radius of convergence R, and let ϕ be defined by

$$\phi(x) = \sum_{n=0}^{\infty} a_n x^n \quad \text{for} \quad -R < x < R.$$

Then ϕ is differentiable on $]-R,R[$ and

$$\phi'(x) = \sum_{n=1}^{\infty} na_n x^{n-1} \quad \text{for} \quad -R < x < R.$$

<u>Proof</u>.   Aim to evaluate $\phi'(c)$ where c is a point
in $]-R,R[$.   Choose a positive number k such that

$c\epsilon$]$-k,k$[<u>and</u> $k < R$.   Let $x\epsilon$]$-k,k$[.   Then

$$\frac{\phi(x) - \phi(c)}{x - c} = \sum_{n=1}^{\infty} a_n \left(\frac{x^n - c^n}{x - c} \right)$$

$$= a_1 + \sum_{n=2}^{\infty} a_n (x^{n-1} + x^{n-2}c + x^{n-3}c^2 + \ldots + c^{n-1}).$$

So $\left| \dfrac{\phi(x) - \phi(c)}{x - c} - \sum_{n=1}^{\infty} n a_n c^{n-1} \right|$

$$= \left| \sum_{n=2}^{\infty} a_n [(x^{n-1} - c^{n-1}) + (x^{n-2}c - c^{n-1}) + \ldots + (xc^{n-2} - c^{n-1})] \right|$$

$$\leq \sum_{n=2}^{\infty} |a_n| [|x^{n-1} - c^{n-1}| + |c||x^{n-2} - c^{n-2}| + \ldots + |c|^{n-2}|x-c|]$$

$$\leq \sum_{n=2}^{\infty} |a_n| [(n-1)k^{n-2}|x-c| + (n-2)k^{n-2}|x-c| + \ldots + 1 \cdot k^{n-2}|x-c|]$$

(estimating as in Ex.1.3)

$$= |x - c| \sum_{n=2}^{\infty} \tfrac{1}{2} n(n-1)|a_n| k^{n-2}$$

$$= |x - c| \cdot S \, . \hspace{4cm} \ldots (*)$$

(Notice that S is finite because $0 < k < R$ and R is the radius of convergence of the second derivative of $\sum a_n x^n$, i.e. of $\sum n(n-1)a_n x^{n-2}$ by two applications of §197.)   Now let $x \to c$ in $(*)$ and deduce that

$$\phi'(c) = \sum_{n=1}^{\infty} n a_n c^{n-1}.$$

<u>Notice</u>: (i) At points actually <u>on</u> the circle of convergence no conclusion about the derivative is possible from this result.   For example, while $\sum x^n/n^2$ is convergent at $x = 1$, its term by term derivative $\sum x^{n-1}/n$ is divergent there.

(ii) It follows from the above result that ϕ is continuous on]$-R,R$[because it is differentiable there.

(iii) The above proof applies with only minor changes to justify complex differentiation at all points z with $|z| < R$.   This is important in Complex Analysis.

<u>EXAMPLE</u>.   By differentiating the binomial expansion of $(1-x)^{-1}$ for $-1 < x < 1$, prove that, for such x, $x + 2x^2 + 3x^3 + 4x^4 + \ldots$ has the sum $x(1-x)^{-2}$.

<u>Solution</u>.   The binomial expansion (valid for $|x| < 1$) is

$$(1-x)^{-1} = 1 + x + x^2 + x^3 + \ldots \, .$$

Differentiate both sides (for $|x| < 1$) using the above result to see that

$$(1 - x)^{-2} = 1 + 2x + 3x^2 + 4x^3 + \ldots .$$

Multiply this by x to get the required result.

EXAMPLES 9

1. For each of the following series find the radius of convergence R and settle whether the series converges for $z = \pm R$:

(a) $\sum \frac{z^n}{n^2}$,   (b) $\sum \frac{2^n z^n}{n^2 + 1}$,   (c) $\sum (-1)^n \frac{2^n z^n}{\sqrt{(n + 1)}}$.

2. Find the radius of convergence of

(a) $\sum \frac{(3n)!}{(n!)^3} z^n$,   (b) $\sum (-1)^n n^2 z^n$,   (c) $\sum \frac{z^n}{(2n)!}$,

(d) $\sum \frac{3.7.11 \ldots (4n - 1)}{2.4.6 \ldots (2n)} z^n$,   (e) $\sum n^n z^n$,   (f) $\sum \frac{4^n z^{2n}}{n^2}$,

(g) $\sum \frac{a(a + 1)(a + 2) \ldots (a + n - 1)}{b(b + 1)(b + 2) \ldots (b + n - 1)} \cdot \frac{z^n}{n}$ $(a, b > 0)$.

3. Show that the radius of convergence of $\sum \frac{(n!)^3}{(3n)!} z^n$ is 27 and by considering the ratio of successive terms show that the series diverges at all points z with $|z| = 27$.

4. Prove that the series $\sum \left(\frac{z - 1}{z + 1}\right)^n$, where $z \varepsilon C - \{-1\}$, converges if and only if $\text{Re } z > 0$. (Notice that this series does <u>not</u> qualify as a <u>power series</u>, so that the points at which it converges need not lie within a circle.)

5. Prove that if $\sum a_n z^n$ and $\sum b_n z^n$ have radii of convergence R_1 and R_2 respectively then $\sum (a_n + b_n) z^n$ has radius of convergence $\min (R_1, R_2)$.

6. Show that the ratio test fails to determine the radius of convergence R of the series

$$2z + 2.3z^2 + 2^2.3z^3 + 2^2.3^2 z^4 + 2^3.3^2 z^5 + \ldots .$$

Show that $R = 1/\sqrt{6}$.

7. Show that the ratio test fails to determine the radius of convergence R of the series

$$2z + 2z^2 + 2^2 z^3 + 2^2 z^4 + 2^3 z^5 + \ldots .$$ Find R.

F

8. Show that if p and q are real polynomial functions and a is a positive number then $\sum \frac{p(n)}{q(n)} a^n z^n$ has radius of convergence $1/a$.

9. The function f is defined on $]0,\infty[$ by $f(x) = x^2/(1+x^3)$. Find integers p and q such that $f(x) = O(x^p)$ as $x \to 0+$ and $f(x) = O(x^q)$ as $x \to \infty$.

10. Find
   (a) $\lim\limits_{x\to 0} \dfrac{\sin ax}{x}$,   (b) $\lim\limits_{x\to 0} \dfrac{1 - \cos ax}{x^2}$,

   (c) $\lim\limits_{x\to 0} \dfrac{2 \sin x - \sin 2x}{x^3}$,   (d) $\lim\limits_{x\to 0} \dfrac{(1-ax)^\alpha - (1+ax)^{-\alpha}}{x^2}$.

11. Show that if $(1+\alpha x)^\beta - (1+\gamma x)^\delta = O(x^3)$ as $x \to 0$, where $\beta\delta \neq 0$, then $\alpha = \gamma$ and $\beta = \delta$.

12. By differentiating the series for $(1 - x)^{-1}$ inside its circle of convergence, prove that

$$\sum_{n=1}^{\infty} n^2 x^n = \frac{x(1 + x)}{(1 - x)^3} \quad \text{for} \quad -1 < x < 1.$$

13. Suppose it is given that $f(x) = \sum\limits_{n=0}^{\infty} a_n x^n$ for $-R < x < R$, and that also $f(x) = b_0 + b_1 x + b_2 x^2 + O(x^3)$ for $0 < x < r$, where $r < R$. Prove that $a_k = b_k$ for $k = 0, 1, 2$.

14. Suppose that the real power series $\sum a_n x^n$ has infinite radius of convergence and further that its sum function ϕ is periodic on R. From Ex.7.5 it is known that all terms of the sequence $\{a_0, a_1, a_2, \ldots\}$ cannot have the same sign. By considering the derivatives of ϕ, prove that the sequence must contain infinitely many terms of each sign.

15. Let $\sum a_n z^n$ be a power series in which the coefficients a_n are all real and with radius of convergence r. Let $b_n = \max(a_n, 0)$, (i.e. $b_n = a_n$ if $a_n > 0$ and $b_n = 0$ otherwise). Prove that the radius of convergence of $\sum b_n z^n$ is at least r. (Hint: Remember that $\sum a_n z^n$ is absolutely convergent inside its circle of convergence.)

X. THE EXPONENTIAL, LOGARITHMIC, TRIGONOMETRIC
AND HYPERBOLIC FUNCTIONS

199. The possibility of defining new functions using power series was mentioned in §188. The exponential function has a special place among all functions so defined, partly because of its own properties, partly because of the properties of its inverse (the logarithmic function) and partly because it acts as a building block for the construction of the sine, cosine, hyperbolic sine and hyperbolic cosine functions.

We look first at the exponential function defined on C (not just on R). The special properties of the <u>real</u> exponential function (i.e. the restriction of the above function to R) then arise naturally.

200. THE EXPONENTIAL FUNCTION

<u>Definition.</u> For every $z\epsilon C$, define

$$\exp z = \sum_{n=0}^{\infty} \frac{z^n}{n!} = 1 + z + \frac{z^2}{2!} + \frac{z^3}{3!} + \ldots .$$

(Notice that in the summation $0! = 1$ and $0^0 = 1$.)

The function exp is then defined on C and is called the <u>exponential function</u>. Notice that the function is indeed well-defined on C, the series being absolutely convergent on the whole of the complex plane: the radius of convergence is infinite (§191) because $(1/(n+1)!)/(1/n!) \rightarrow 0$ as $n \rightarrow \infty$.

Notice that $\exp 0 = 1$ and that

$$\exp 1 = 1 + 1 + 1/2! + 1/3! + \ldots = 2.71828\ldots .$$

This number, exp 1, is called e.

201. THE FUNCTIONAL RELATION FOR THE EXPONENTIAL FUNCTION

There is no apparent reason why the function defined as the sum function of $\sum z^n/n!$ (i.e. the exponential function) should merit more attention than the sum functions of say $\sum z^n/n^n$ or $\sum z^n/n^2$. The special quality that distinguishes the exponential function from all others defined by power series derives from the property established in the result below. This property arises because the terms generated by multiplying together the series for exp a and exp b just happen to combine to give the series for $\exp(a+b)$. Attempts to repeat the trick with other power series like $\sum z^n/n^n$ and $\sum z^n/n^2$ fail.

<u>RESULT</u>.   For all   a, b ε C,

$$\exp a . \exp b = \exp(a + b).$$

<u>Proof</u>.   Multiplying together two absolutely convergent series using the result of §186, we get

$$\exp a . \exp b = \sum_{r=0}^{\infty} \frac{a^r}{r!} \sum_{s=0}^{\infty} \frac{b^s}{s!}$$

$$= \sum_{r=0}^{\infty} \sum_{s=0}^{\infty} \frac{a^r b^s}{r! s!} . \qquad \ldots (*)$$

In (*) collect together all terms of combined degree   n (n $\geq$ 0);   i.e. let   r + s = n.   So the general term in the product is c_n,   where

$$c_n = \sum_{r=0}^{n} \frac{a^r}{r!} \cdot \frac{b^{n-r}}{(n-r)!}$$

$$= \frac{1}{n!} \sum_{r=0}^{n} \frac{n!}{r!(n-r)!} a^r b^{n-r}$$

$$= \frac{1}{n!} \sum_{r=0}^{n} \binom{n}{r} a^r b^{n-r} = \frac{(a+b)^n}{n!} .$$

So from (*) we can conclude that

$$\exp a . \exp b = \sum_{n=0}^{\infty} \frac{(a+b)^n}{n!} = \exp(a + b).$$

202.   WHY EXP α IS e^{α} (FOR THE CASE OF RATIONAL α)

From §201 notice that, for example, $\exp 2 = \exp 1 . \exp 1 = (\exp 1)^2 = e^2$.   It is easy to extend this to show that $\exp n = e^n$ for every nεN. Also, since   exp n . exp (-n) = exp (n - n) = exp 0 = 1, conclude that $\exp (-n) = 1/\exp n = e^{-n}$ for every   nεN.

We can then go further and notice that

$$\exp \tfrac{1}{2} . \exp \tfrac{1}{2} = \exp 1 = e.$$

So   exp $\tfrac{1}{2}$ = $\pm\sqrt{e}$.   But the negative sign can be discarded because it is obvious from the series definition of exp $\tfrac{1}{2}$ that   exp $\tfrac{1}{2}$ > 0.   So we find that   exp $\tfrac{1}{2}$ = $\sqrt{e}$. In a similar way we can show that, for every <u>rational</u> number α,   exp α (thought of as the sum of the series) turns out to coincide with e^{α} (as defined in §90).

203. THE MEANING OF e^α FOR IRRATIONAL AND COMPLEX α

Notice that <u>irrational</u> and <u>complex</u> powers like $3^{\sqrt{2}}$ and 6^{1+i} have not been defined yet. As a first step on this road we show how to define <u>irrational</u> and <u>complex</u> powers of the number e, for example $e^{\sqrt{2}}$ and e^{1+i}. From §202 remember that, for a <u>rational</u> number α, $e^\alpha = \exp \alpha$. By analogy, for an <u>irrational</u> or <u>complex</u> power β, <u>define</u>

$$e^\beta = \exp \beta.$$

(Here the series definition for $\exp \beta$ makes sense even for irrational and complex β.)

The functional relation for the exponential (§201) guarantees that such powers obey the first index law, $e^\alpha e^\beta = e^{\alpha+\beta}$, for all real and complex α and β. The second index law, $(e^\alpha)^\beta = e^{\alpha\beta}$ is also true in the case when α and β are both real (§210) but it can fail if α and β are both complex (§228).

204. COMPLEX TRIGONOMETRIC FUNCTIONS

Notice from §200 and §203 that, for all $z \varepsilon C$,

$$e^{iz} = 1 + iz - \frac{z^2}{2!} - \frac{iz^3}{3!} + \ldots$$

and

$$e^{-iz} = 1 - iz - \frac{z^2}{2!} + \frac{iz^3}{3!} + \ldots \ .$$

Then <u>define</u> $\cos z$ and $\sin z$ for every $z \varepsilon C$ by

$$\cos z = \tfrac{1}{2}(e^{iz} + e^{-iz}) = 1 - \frac{z^2}{2!} + \frac{z^4}{4!} - \frac{z^6}{6!} + \ldots \ ,$$

$$\sin z = \frac{1}{2i}(e^{iz} - e^{-iz}) = z - \frac{z^3}{3!} + \frac{z^5}{5!} - \frac{z^7}{7!} + \ldots \ .$$

<u>Notice</u>: (i) The series for $\cos z$ and $\sin z$ are absolutely convergent for all $z \varepsilon C$.

(ii) $\cos 0 = 1$ and $\sin 0 = 0$. Also $\cos$ is an <u>even</u> function and $\sin$ is an <u>odd</u> function, i.e. $\cos(-z) = \cos z$ and $\sin(-z) = -\sin z$. The respective power series contain only <u>even</u> and <u>odd</u> powers of z in line with Ex.7.1.

(iii) $\tan z$, $\sec z$, $\csc z$ and $\cot z$ are defined as $\sin z/\cos z$, $1/\cos z$, $1/\sin z$ and $1/\tan z$ respectively, wherever the denominators are non-zero.

(iv) Identities like

$$\cos(a + b) = \cos a \cos b - \sin a \sin b$$

and

$$\sin(a + b) = \sin a \cos b + \cos a \sin b$$

follow easily from the definitions.   For example,

$$\cos a \cos b - \sin a \sin b = \frac{1}{4}\left[(e^{ia}+e^{-ia})(e^{ib}+e^{-ib})\right.$$
$$\left. + (e^{ia}-e^{-ia})(e^{ib}-e^{-ib})\right]$$
$$= \tfrac{1}{2}\left(e^{i(a+b)} + e^{-i(a+b)}\right) \quad \text{(on simplifying)}$$
$$= \cos(a+b).$$

Putting $b = -a$ in this identity gives the result that

$$\cos^2 z + \sin^2 z = 1 \quad \text{for all} \quad z\varepsilon C.$$

EXAMPLE.   Prove that $\sin 2z = 2\sin z \cos z$ and that $\cos 2z = 2\cos^2 z - 1$ from the above definitions.

205.   COMPLEX HYPERBOLIC FUNCTIONS

Notice from §200 and §203 that, for all $z\varepsilon C$,

$$e^z = 1 + z + \frac{z^2}{2!} + \frac{z^3}{3!} + \dots$$

and $\quad e^{-z} = 1 - z + \frac{z^2}{2!} - \frac{z^3}{3!} + \dots \;.$

Then define the hyperbolic cosine and hyperbolic sine functions for every $z\varepsilon C$ by

$$\cosh z = \tfrac{1}{2}(e^z + e^{-z}) = 1 + \frac{z^2}{2!} + \frac{z^4}{4!} + \frac{z^6}{6!} + \dots$$

and $\quad \sinh z = \tfrac{1}{2}(e^z - e^{-z}) = z + \frac{z^3}{3!} + \frac{z^5}{5!} + \frac{z^7}{7!} + \dots \;.$

Notice: (i)   These series are absolutely convergent at all points of C.

(ii) $\cosh 0 = 1$ and $\sinh 0 = 0$.   Also cosh is even and sinh is odd and this is reflected in the even and odd powers in the respective expansions.

(iii)   Identities like $\cosh^2 z - \sinh^2 z = 1$ follow from the definitions in terms of exponentials.

(iv)   tanh z, sech z, cosech z and coth z are defined as sinh z/cosh z, 1/cosh z, 1/sinh z and 1/tanh z respectively, wherever the denominators are non-zero.

(v)   From the non-series forms of the definitions it is easy to show that

$$\sin iz = i \sinh z, \qquad \cos iz = \cosh z,$$
$$\sinh iz = i \sin z, \qquad \cosh iz = \cos z.$$

EXAMPLE.   Prove that $\sinh 2z = 2\sinh z \cosh z$ and that $\cosh 2z = \cosh^2 z + \sinh^2 z$ from the above definitions.

REAL EXPONENTIAL AND LOGARITHMIC FUNCTIONS

PROPERTIES OF THE REAL EXPONENTIAL FUNCTION

The real exponential function is defined for every $x \varepsilon R$ by

$$\exp x = \sum_{n=0}^{\infty} \frac{x^n}{n!} \, ,$$

with the conventions that $0! = 1$ and $0^0 = 1$.

This is just the restriction of the exponential function defined on C in §200. In the light of §202 (a matter of fact) and §203 (a matter of definition), we can choose to write e^x instead of $\exp x$.

RESULT. The function e^x is positive, strictly increasing and differentiable on R. Furthermore, $e^x \to \infty$ as $x \to \infty$, $e^x \to 0$ as $x \to -\infty$ and $\frac{d}{dx}(e^x) = e^x$ for all $x \varepsilon R$.

Proof. Taking the series as $e^x = 1 + x + \frac{x^2}{2!} + \frac{x^3}{3!} + \dots$,

notice that $e^0 = 1$ and that, for $x > 0$, all the terms are positive so that $e^x > 1$. Also, from the result of §201, notice that $e^{-x}.e^x = e^0 = 1$: so $e^{-x} = 1/e^x$ and $0 < e^y < 1$ for all $y < 0$.

Furthermore, since the radius of convergence is infinite, it follows from §198 that e^x is differentiable (and so continuous) on R. Also

$$\frac{d}{dx}(e^x) = \frac{d}{dx}(1 + x + \frac{x^2}{2!} + \frac{x^3}{3!} + \dots) = 1 + x + \frac{x^2}{2!} + \frac{x^3}{3!} + \dots = e^x .$$

But then since $e^x > 0$ for all $x \varepsilon R$ (as shown above) it follows that the derivative of e^x is positive on R: so e^x is strictly increasing on R by §130. Moreover, from the series it follows that $e^x > 1 + x$ for all $x > 0$. So $e^x \to \infty$ as $x \to \infty$. Also,

$$\lim_{x \to -\infty} e^x = \lim_{y \to \infty} e^{-y} = \lim_{y \to \infty} 1/e^y = 0.$$

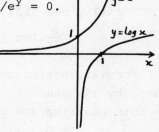

Using the above result we can draw the graph of e^x and that of its inverse $\log x$.

EXAMPLE. Notice that by composition of continuous functions we can say, for example, that

$e^{x/(x^2+1)} \to e^0 = 1$ as $x \to \infty$ and $e^{-1/x^2} \to 0$ as $x \to 0$.

207. DEFINITION AND PROPERTIES OF LOG X FOR POSITIVE VALUES OF X

Recall from §89 and §123 that a strictly increasing function f (like e^x) brings with it a corresponding inverse function f^{-1} and that properties of f are inherited by f^{-1}. Accordingly we have the following result.

RESULT. The real exponential function has a unique inverse function, the logarithmic function, denoted by log and with the following properties.

(1) $\log : \,]0,\infty[\,\to R$ is strictly increasing, continuous and differentiable.

(2) $\log(e^x) = x$ and $e^{\log y} = y$ for all $x \varepsilon R$ and all $y > 0$.

(3) $\log y = x \iff y = e^x$, for $y > 0$ and $x \varepsilon R$.

(4) $\log 1 = 0$ and $\log e = 1$.

(5) $\log(ab) = \log a + \log b$ for all $a, b > 0$.

(6) $\log(1/a) = -\log a$ for all $a > 0$.

(7) $\frac{d}{dy}(\log y) = \frac{1}{y}$ for all $y > 0$.

(8) $\log y \to \infty$ as $y \to \infty$ and $\log y \to -\infty$ as $y \to 0+$.

Proof. Apply the results on inverse functions from §89 and §123 with $f(x) = e^x$. Parts (1),(2),(3) and (4) are then immediate.

For (5), let a, b be positive numbers and let $a = e^c$ and $b = e^d$. Then $ab = e^c.e^d = e^{c+d}$. So $\log(ab) = c + d = \log a + \log b$.

For (6), put $b = 1/a$ in (5) and use (4).

For (7), use §123 to see that if $y = e^x$ then

$$\frac{d}{dy}(\log y) = \frac{1}{\frac{d}{dx}(e^x)} = \frac{1}{e^x} = \frac{1}{y}.$$

For (8), notice that log is strictly increasing and that, by (5) and (6), $\log 2^n = n \log 2$ for all integers n both positive and negative.

EXAMPLE. Show that $\log((x^2+1)/x^2)$ tends to infinity as $x \to 0$ and tends to zero as $x \to \infty$.

208. ALTERNATIVE DEVELOPMENT OF THE EXPONENTIAL AND LOGARITHMIC FUNCTIONS

Every Analysis course has to develop the exponential and logarithmic functions somehow. In this account the exponential is developed as the sum function of a power series. This demands knowledge of where power series converge (to show exp is well-defined on C), of where their sum functions are differentiable (for the derivative) and of how to multiply them together (for the functional relation). The logarithmic function then arises as the inverse of the exponential.

Another approach is through integration. Once the theory of integration is established the logarithmic function can be defined, for $x > 0$, by

$$\log x = \int_1^x \frac{dt}{t} \ .$$

A result from the theory of integration (§245) then guarantees that $\frac{d}{dx}(\log x) = \frac{1}{x}$. Further, by writing

$$\int_1^{xy} \frac{dt}{t} = \int_1^x \frac{dt}{t} + \int_x^{xy} \frac{dt}{t} \qquad \ldots (*)$$

and changing the variable in the third integral it is possible to show that $\log(xy) = \log x + \log y$. The real exponential function then arises as the inverse of the logarithmic function. The derivative of the exponential is then found from the derivative of log x using the result of §123, and this leads on to finding the Maclaurin series for the exponential. The functional relation of §201 can then be recovered from the functional relation for the logarithm proved in (*) above.

209. THE HIERARCHY: EXP X , POWERS OF X , LOG X .

In general terms the following result says that in a competitive situation e^x dominates any fixed positive power of x and that any fixed positive power of x dominates log x.

RESULT. every number $k > 0$,

$$e^x/x^k \to \infty \qquad \text{as} \quad x \to \infty,$$

$$x^k/(\log x) \to \infty \qquad \text{as} \quad x \to \infty,$$

$$x^k \log x \to 0 \qquad \text{as} \quad x \to 0+.$$

Proof.   (1)   Take $p \varepsilon N$ with $p > k+1$.   Then,

$$e^x/x^k = (1 + x + \frac{x^2}{2!} + \ldots + \frac{x^p}{p!} + \ldots)/x^k > \frac{x^{p-k}}{p!} \quad \text{for} \quad x > 0$$

$$\rightarrow \infty \quad \text{as} \quad x \rightarrow \infty.$$

(2)   Take $x > 0$ and put $x = e^y$.   So,

$$\frac{x^k}{\log x} = \frac{(e^y)^k}{\log e^y} = \frac{e^{ky}}{y} \rightarrow \infty \quad \text{as} \quad y \rightarrow \infty \text{ by (1), i.e. as } x \rightarrow \infty.$$

(3)   Put $x = 1/y$.   So,

$$x^k \log x = y^{-k} \log (1/y) = -y^{-k} \log y \rightarrow 0$$

$$\text{as} \quad y \rightarrow \infty \text{ by (2), i.e. as } x \rightarrow 0+.$$

(Notice that the second index law used in the proof of (2) is justified in §210.)

EXAMPLE. . Show that there exists a number   N   such that
$$(\log x)^4 < x \quad \text{for all} \quad x \geq N.$$

Solution.   By the hierarchy result above, $x^{1/4}/(\log x) \rightarrow \infty$
as $x \rightarrow \infty$.   So $x/(\log x)^4 \rightarrow \infty$ as $x \rightarrow \infty$.   So
$\exists$ N   such that $x/(\log x)^4 > 1$ for all $x \geq N$,
i.e. $(\log x)^4 < x$ for all $x \geq N$.

EXAMPLE.   Discuss the behaviour as $x \rightarrow \infty$ of
$$\text{(a)} \quad (\log x)^4 x^5 e^{-2x}, \quad \text{(b)} \quad x^6 e^{-(\log x)^2}.$$

Solution.   (a) Use the last example to find   N   such that
$$0 < (\log x)^4 x^5 e^{-2x} < x^6 e^{-2x} = (x^3 e^{-x})^2 \quad \forall x \geq N.$$
Here the final term tends to zero by the hierarchy result.   So by a sandwich argument we conclude that (a) tends to zero as $x \rightarrow \infty$.

(b)   Take $x > 0$ and put $y = \log x$.   Then,
$$\lim_{x \rightarrow \infty} x^6 e^{-(\log x)^2} = \lim_{y \rightarrow \infty} e^{6y} e^{-y^2} = \lim_{y \rightarrow \infty} e^{-y^2 + 6y} = 0,$$
because $-y^2 + 6y \rightarrow -\infty$ as $y \rightarrow \infty$.

EXAMPLE.   Prove that $\sum \frac{(\log n)^4}{n^2}$ is convergent.

<u>Solution</u>. Proceed along the lines of the first example to find M with $(\log n)^4 < \sqrt{n}$ for all $n \geq M$.   Then,

$$\frac{(\log n)^4}{n^2} < \frac{\sqrt{n}}{n^2} = \frac{1}{n^{3/2}} \quad \forall n \geq M,$$

so that the given series converges by comparison.

[A free translation of part (2) of the hierarchy result is that "log n is <u>meaner</u> than every positive power of n".   This is the thinking behind this last solution.   See also §215.]

210. IRRATIONAL POWERS OF POSITIVE REAL NUMBERS

The need for a definition of these (e.g. $3^{\sqrt{2}}$) was expressed in §203.   There we <u>defined</u> $e^\beta = \exp \beta$ for all $\beta \epsilon R$.   The following definition extends this.

<u>Definition</u>.   Let $a > 0$.   Then, for every $x \epsilon R$, define a^x by $a^x = e^{x \log a}$.

(Notice also that $0^x = 0$ for all $x > 0$ and that $0^0 = 1$ by convention.)

It is routine to check that this definition does not conflict with our earlier ideas about rational powers. For example, $8^{2/3} = \exp(\frac{2}{3} \log 8) = 4$.

The first index law, $a^x a^y = a^{x+y}$ for $a > 0$ and $x, y \epsilon R$, follows directly from the functional relation for the exponential in §201.   To check the second index law, $(a^x)^y = a^{xy}$ for $a > 0$ and $x, y \epsilon R$, notice that

$$(a^x)^y = \exp(y(\log a^x)) = \exp(y(\log \exp (x \log a)))$$

$$= \exp(y(x \log a)) = \exp(xy \log a) = a^{xy}, \text{ as required.}$$

This second index law fails for complex powers (§228).

211. TWO EXAMPLES ON FRACTIONAL POWERS

Apart altogether from its theoretical importance, the fact that $a^x = e^{x \log a}$ is very useful in practice.

<u>EXAMPLE</u>.   Show that $n^{1/n}$, $n^{1/\sqrt{n}}$ and $a^{1/n}$ $(a > 0)$ all tend to 1 as $n \to \infty$.

<u>Solution</u>.   Write the candidates as $\exp\left(\frac{1}{n} \log n\right),$

$\exp\left(\frac{1}{\sqrt{n}} \log n\right)$ and $\exp\left(\frac{1}{n} \log a\right)$. In each case the bracket tends to zero as $n \to \infty$, clearly in the last case and by the hierarchy result of §209 in the first two cases. So each candidate tends to $\exp 0 = 1$.

(Contrast the flexibility of this method with the method of §92.)

<u>EXAMPLE</u>. Find the limit as $n \to \infty$ of $n(2^{1/n} - 1)$.

<u>Solution</u>. $n(2^{1/n} - 1) = n\left(\exp\left(\frac{1}{n} \log 2\right) - 1\right)$

$= n\left(1 + \left(\frac{1}{n} \log 2\right) + O\left(\left(\frac{1}{n} \log 2\right)^2\right) - 1\right)$ for all $n \geq N$, say,

on using the series for $\exp$ and O-notation (§265, §193).

$= \log 2 + O\left(\frac{1}{n}\right) \to \log 2$ as $n \to \infty$.

For further examples of these types see Ex.10.6 parts (c), (d), (e), (f).

212. THE MACLAURIN SERIES FOR $\log(1 + X)$

Since $\log 0$ is undefined there is no Maclaurin series for $\log x$. On the other hand, $\log x$ has infinitely many derivatives at $x = 1$ and so there is a Taylor series for $\log x$ round $x = 1$, i.e. there is a Maclaurin series for $\log(1 + x)$.

<u>RESULT</u>. For $-1 < x \leq 1$,

$$\log(1 + x) = x - \frac{x^2}{2} + \frac{x^3}{3} - \frac{x^4}{4} + \ldots = \sum_{n=1}^{\infty} (-1)^{n-1} \frac{x^n}{n}.$$

<u>Proof</u>. Let $f(x) = \log(1 + x)$. Then $f'(x) = (1 + x)^{-1}$ and in general $f^{(n)}(x) = (-1)^{n-1}(n-1)!\,(1 + x)^{-n}$. So $f(0) = 0$, $f'(0) = 1$, ..., $f^{(n)}(0) = (-1)^{n-1}(n-1)!$. By Taylor's theorem,

$$\log(1 + x) = x - \frac{x^2}{2} + \frac{x^3}{3} - \ldots + \frac{(-1)^{n-2} x^{n-1}}{(n - 1)} + R_n(x).$$

The expansion as an infinite series will be established for those values of x for which $R_n(x) \to 0$ as $n \to \infty$. For $0 \leq x \leq 1$, with Lagrange's remainder,

$$|R_n(x)| = \frac{x^n}{n(1 + \xi)^n}, \text{ where } 0 < \xi < 1,$$

and since $x \leq 1$ and $(1 + \xi) > 1$, it follows that
$|R_n(x)| \leq \frac{1}{n}$ and so $|R_n(x)| \to 0$ as $n \to \infty$ for such x.
The result for $x\varepsilon[0,1]$ then follows.

For $x\varepsilon]-1,0[$, use Cauchy's remainder in a slightly more delicate argument similar to that in §174.

Notice: (i) The result does not hold for $x = -1$ since log 0 is undefined.

(ii) The fact that $x \leq 1$ is vital in getting the remainder to tend to zero in the above argument.

(iii) Putting $x = 1$ in the above series shows that the logarithmic series (§155 and §181) converges to log 2, i.e. $1 - \frac{1}{2} + \frac{1}{3} - \frac{1}{4} + \frac{1}{5} - \ldots = \log 2 = 0.6931\ldots$.

213. TWO OTHER MACLAURIN SERIES

These are derived from the series for $\log(1 + x)$.

RESULT. $\log(1 - x) = -x - \frac{x^2}{2} - \frac{x^3}{3} - \ldots$ for $x\varepsilon[-1,1[$,

and $\log\left(\frac{1 + x}{1 - x}\right) = 2(x + \frac{x^3}{3} + \frac{x^5}{5} + \ldots)$ for $x\varepsilon]-1,1[$.

Proof. For the first, replace x by $-x$ in the series for $\log(1 + x)$. For the second, subtract the first from the series for $\log(1 + x)$.

EXAMPLE. Show (a) $\sum_{n=1}^{\infty} \frac{1}{n} \left(\frac{1}{2^n} - \frac{1}{3^n} - \frac{1}{4^n}\right) = 0$,

(b) $\frac{1}{3} + \frac{1}{3} \cdot \frac{1}{3^3} + \frac{1}{5} \cdot \frac{1}{3^5} + \ldots = \frac{1}{2}\log 2$.

214. AN IMPORTANT LIMIT

Both the following result and its method of proof deserve attention.

RESULT. For every $a\varepsilon R$, $\left(1 + \frac{a}{n}\right)^n \to e^a$ as $n \to \infty$.

In particular, $\left(1 + \frac{1}{n}\right)^n \to e$ as $n \to \infty$.

Proof. (Compare §211.) $\left(1 + \frac{a}{n}\right)^n = \exp\left(n \log\left(1 + \frac{a}{n}\right)\right) =$

$$= \exp\left(n\left[\frac{a}{n} + O\left(\left[\frac{a}{n}\right]^2\right)\right]\right) \quad \text{for all} \quad n > N, \text{ say, on using the series for} \quad \log(1+x)$$

$$= \exp\left(a + O\left[\frac{1}{n}\right]\right) \to e^a \quad \text{as} \quad n \to \infty.$$

(Here the series for $\log(1+x)$ is used to expand $\log\left(1 + \frac{a}{n}\right)$. This is justified for sufficiently large n because then $\left|\frac{a}{n}\right| < 1$.)

EXAMPLE. Show that $\left(\frac{n+2}{n-2}\right)^n \to e^4$ and $\left(1 + \frac{1}{n^2}\right)^n \to 1$ as $n \to \infty$. (Hint: Divide above and below by n inside the bracket in the first. Use the method of the above result for the second.)

215. THE DIVERGENT SERIES $\sum 1/(n \log n)$

This series is important as a source of counter-examples. It is divergent by the integral test. More generally the integral test tells us that the series

$$\sum \frac{1}{n(\log n)^k}$$ converges for $k > 1$ and diverges for $k \leq 1$.

Using the method of §179, we can show that

$$\frac{1}{2 \log 2} + \frac{1}{3 \log 3} + \cdots + \frac{1}{n \log n} = \log\log n + c + \alpha_n,$$

where c is a constant and $\alpha_n \to 0$ as $n \to \infty$.

In the wake of the hierarchy result of §209 we remarked that in general terms $\log n$ is "meaner" than every positive power of n. Further evidence of this meanness is that though $\sum \frac{1}{n^{1+k}}$ converges for every

value of $k > 0$, $\sum \frac{1}{n \log n}$ still diverges.

Notice finally that $\sum \frac{1}{n}$, $\sum \frac{1}{n \log n}$,

$\sum \frac{1}{n(\log n)(\log\log n)}$, ... are all divergent.

THE REAL TRIGONOMETRIC AND INVERSE TRIGONOMETRIC FUNCTIONS

216. THE COSINE AND SINE FUNCTIONS

Taking the definitions of $\cos x$ and $\sin x$ ($x \varepsilon R$) as particular cases of the definitions in §204, notice that, for all $x \varepsilon R$,

$$\cos x = 1 - \frac{x^2}{2!} + \frac{x^4}{4!} - \cdots \quad \text{and} \quad \sin x = x - \frac{x^3}{3!} + \frac{x^5}{5!} - \cdots .$$

Notice also from §204 that $\cos^2 x + \sin^2 x = 1$ and that $\cos(x + y) = \cos x \cos y - \sin x \sin y$.

It follows from the result on differentiation of power series (§198) that $\cos x$ and $\sin x$ are differentiable and therefore continuous on R. Also we see on term by term differentiation that

$$\frac{d}{dx}(\cos x) = -\sin x \quad \text{and} \quad \frac{d}{dx}(\sin x) = \cos x.$$

<u>EXAMPLE</u>. Use truncation of power series to show that

$$\lim_{x \to 0} \frac{1 - \cos x}{x^2} = \tfrac{1}{2} \quad \text{and} \quad \lim_{x \to 0} \frac{\cos ax - \cos bx}{x^2} = \tfrac{1}{2}(b^2 - a^2).$$

217. THE NUMBER π

The following result shows how the number π enters the trigonometric arena set by §216. Think of trying to draw the graphs of $\cos x$ and $\sin x$ $(x \varepsilon R)$ starting from the definitions in §216.

<u>RESULT</u>. The cosine function (as defined in §216) does take the value zero on $]0,\infty[$.

<u>Proof</u>. Notice first that, since $\cos^2 x + \sin^2 x = 1$, $-1 \le \cos x \le 1$ and $-1 \le \sin x \le 1$ for all $x \varepsilon R$.

Suppose it were the case that $\cos x$ <u>never</u> took the value 0 on $]0,\infty[$. Then, since $\cos 0 = 1$ and since $\cos$ is continuous, it would follow from the intermediate value theorem (§87) that $\cos x$ would be positive on the whole of $]0,\infty[$, i.e. $\sin x$ would have positive derivative on $]0,\infty[$. So $\sin x$ would be strictly increasing on $[0,\infty[$. But $\sin 0 = 0$ and $\sin x$ is bounded above by 1. So $\sin x$ would tend to a <u>positive</u> limit L as $x \to \infty$. But $-\sin x$ is the derivative of $\cos x$ and so the derivative of $\cos x$ would tend to the <u>negative</u> number $-L$ as $x \to \infty$. So, by an argument similar to §133(i), deduce that $\cos x$ would tend to $-\infty$ as $x \to \infty$. This would contradict the fact that $-1 \le \cos x \le 1$. So the above supposition is wrong. So there does exist at least one point $c \varepsilon]0,\infty[$ such that $\cos c = 0$.

To get at the number π notice from Ex.3.16 that

since cos x is continuous, the set of all positive numbers x such that cos x = 0 must contain its infimum. This infimum is 1.5707963... and we call this number $\frac{1}{2}\pi$. So π = 3.14159 (5 decimal places).

It follows that sin $\frac{1}{2}\pi$ = 1, because cos $\frac{1}{2}\pi$ = 0, $\cos^2 x + \sin^2 x = 1$ and sin x is strictly increasing on $]0, \frac{1}{2}\pi[$. Using the relations that sin 2x = 2 sin x cos x and cos 2x = $\cos^2 x - \sin^2 x$, deduce that cos π = -1, sin π = 0 and further that cos 2π = 1 and sin 2π = 0.

Then again from the addition relations for sin and cos (§204) deduce that

cos(x + 2π) = cos x cos 2π - sin x sin 2π = cos x ,

sin(x + 2π) = sin x cos 2π + cos x sin 2π = sin x .

These relations show that sin and cos are periodic with period 2π. We can then draw the graphs of sin x and cos x as shown. The relationship between π and the area of a circle can be established by integration.

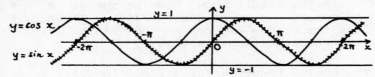

218. The properties of the other trigonometric functions tan, sec, cosec and cot defined in §204 follow from those of sin and cos. Notice that tan is defined at all points of R that are not equal to an odd multiple of $\frac{1}{2}\pi$. Also tan is periodic with period π and tan x $\to \pm\infty$ as x $\to \pm\frac{1}{2}\pi$. Also,

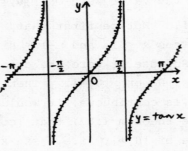

$$\frac{d}{dx}(\tan x) = \sec^2 x, \qquad \frac{d}{dx}(\cot x) = -\mathrm{cosec}^2 x,$$

$$\frac{d}{dx}(\sec x) = \sec x \tan x, \qquad \frac{d}{dx}(\mathrm{cosec}\ x) = -\mathrm{cosec}\ x \cot x.$$

<u>EXAMPLE</u>. Use the addition formulae for sin and cos to show that $\tan(a + b) = \dfrac{\tan a + \tan b}{1 - \tan a \tan b}$ (a, b ϵ C).

219. INVERSE TRIGONOMETRIC FUNCTIONS

The results of §89 and §123 on inverse functions can be applied to bijective restrictions of sin, cos and tan. The standard inverses are shown in the table on page 169.

Function (f)	sin x	cos x	tan x
Inverse (f^{-1})	$\sin^{-1}y$	$\cos^{-1}y$	$\tan^{-1}y$
Domain of f^{-1}	$[-1,1]$	$[-1,1]$	$]-\infty,\infty[$
Range of f^{-1}	$[-\tfrac{1}{2}\pi,\tfrac{1}{2}\pi]$	$[0,\pi]$	$]-\tfrac{1}{2}\pi,\tfrac{1}{2}\pi[$
$\dfrac{d}{dy}(f^{-1}(y))$	$\dfrac{1}{\sqrt{(1-y^2)}}$	$\dfrac{-1}{\sqrt{(1-y^2)}}$	$\dfrac{1}{1+y^2}$

There is an element of arbitrariness in the choice of the ranges of the inverse functions f^{-1}. Notice that $\tan^{-1}y \to \tfrac{1}{2}\pi$ as $y \to \infty$ and $\tan^{-1}y \to -\tfrac{1}{2}\pi$ as $y \to -\infty$.

§125 develops the derivative of $\sin^{-1}$. To get the derivative of $\tan^{-1}$ use §123 as follows:

$$\frac{d}{dy}(\tan^{-1}y) = \frac{1}{\dfrac{d}{dx}(\tan x)} = \frac{1}{\sec^2 x} = \frac{1}{1+\tan^2 x} = \frac{1}{1+y^2}\,.$$

EXAMPLE.   Use the example of §218 to show that

$$\tan^{-1}x + \tan^{-1}y = \tan^{-1}\left(\frac{x+y}{1-xy}\right) + k\pi,$$

where $k = -1$, 0 or 1.   Illustrate with examples that all three values of k are possible.

220.   THE MACLAURIN SERIES FOR THE INVERSE TANGENT, GREGORY'S SERIES

For $f(x) = \tan^{-1}x$, notice that $(1+x^2)f'(x) = 1$. Differentiating this n times with Leibniz's theorem (as in the example in §178) gives the relation

$$f^{(n+1)}(0) = -n(n-1)f^{(n-1)}(0) \qquad (n\varepsilon N).$$

Starting from $f(0) = 0$ and $f'(0) = 1$, deduce that, for all $n\varepsilon N$,

$$f^{(2n)}(0) = 0 \quad \text{and} \quad f^{(2n-1)}(0) = (-1)^{n-1}(2n-2)!\,.$$

So, the Maclaurin series of $\tan^{-1}x$ is developed as

$$x - \frac{x^3}{3} + \frac{x^5}{5} - \frac{x^7}{7} + \dots \,.$$

As §178 attempts to make clear, this method gives no guarantee that the Maclaurin series converges to $\tan^{-1}x$. Actually it does for $-1 \le x \le 1$: the integration method of developing the series (§247) proves this.

Putting $x = 1$ in this Maclaurin series gives Gregory's series

$$1 - \frac{1}{3} + \frac{1}{5} - \frac{1}{7} + \ldots = \tan^{-1}1 = \frac{\pi}{4}.$$

The convergence of this series is too slow to allow easy calculation of π. Much better are the rapidly convergent series for $\tan^{-1}\frac{1}{5}$ etc in relations like Machin's formula, namely,

$$\frac{\pi}{4} = 4\tan^{-1}\frac{1}{5} - \tan^{-1}\frac{1}{239}.$$

See also Ex.13.14.

THE REAL HYPERBOLIC AND INVERSE HYPERBOLIC FUNCTIONS

221. Taking the definitions of $\cosh x$ and $\sinh x$, where $x \varepsilon R$, as particular cases of the definitions in §205, notice that, for all $x \varepsilon R$,

$$\cosh x = \tfrac{1}{2}(e^{x} + e^{-x}) = 1 + \frac{x^2}{2!} + \frac{x^4}{4!} + \frac{x^6}{6!} + \ldots ,$$

$$\sinh x = \tfrac{1}{2}(e^{x} - e^{-x}) = x + \frac{x^3}{3!} + \frac{x^5}{5!} + \frac{x^7}{7!} + \ldots .$$

Notice also that the identities and other definitions of §205 apply in R as a particular case.

The differentiability and continuity of $\cosh x$ and $\sinh x$ for all $x \varepsilon R$ follow from the differentiability of e^{x} and e^{-x}. Also

$$\frac{d}{dx}(\cosh x) = \sinh x \quad \text{and} \quad \frac{d}{dx}(\sinh x) = \cosh x.$$

Notice also that $\cosh$ is even, $\sinh$ is odd and that $\cosh x \geq 1$ for all $x \varepsilon R$.

If the definition of $\tanh x$ is taken in the form

$$\tanh x = \frac{e^{2x} - 1}{e^{2x} + 1},$$

then it is clear that $\tanh x \to \pm 1$ as $x \to \pm\infty$. Also $\frac{d}{dx}(\tanh x) = \mathrm{sech}^2 x > 0$, so that $\tanh x$ is strictly increasing on R. Here are the graphs.

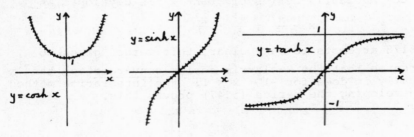

222. INVERSE HYPERBOLIC FUNCTIONS

To develop these apply the results of §89 and §123. Notice that sinh x and tanh x are strictly increasing on R and that cosh x is strictly increasing on $[0,\infty[$. The standard inverses are shown in the following table.

Function (f)	sinh x	cosh x	tanh x
Inverse (f^{-1})	$\sinh^{-1}y$	$\cosh^{-1}y$	$\tanh^{-1}y$
Domain of f^{-1}	R	$[1,\infty[$	$]-1,1[$
Range of f^{-1}	R	$[0,\infty[$	R
$\frac{d}{dy}(f^{-1}(y))$	$\dfrac{1}{\sqrt{(y^2+1)}}$	$\dfrac{1}{\sqrt{(y^2-1)}}$	$\dfrac{1}{1-y^2}$

Notice that to get the derivative of $\sinh^{-1}y$, for example, we apply §123 as follows:

$$\frac{d}{dy}(\sinh^{-1}y) = \frac{1}{\frac{d}{dx}(\sinh x)} = \frac{1}{\cosh x} = \frac{1}{\sqrt{(1+\sinh^2 x)}} = \frac{1}{\sqrt{(y^2+1)}}.$$

Actually these inverse functions are known functions in disguise. For

$$\sinh^{-1}y = \log(y+\sqrt{(y^2+1)}),$$
$$\cosh^{-1}y = \log(y+\sqrt{(y^2-1)}),$$
$$\tanh^{-1}y = \tfrac{1}{2}\log\left(\frac{1+y}{1-y}\right).$$

For example, suppose that $x = \sinh^{-1}y$. Then $y = \sinh x$. So $2y = e^x - e^{-x}$. So $e^{2x} - 2ye^x - 1 = 0$. Treating this equation as a quadratic in e^x find that $e^x = y \pm \sqrt{(y^2+1)}$. Discard the negative sign because e^x must be positive. Then take the logarithm of each side to solve for x. So $\sinh^{-1}y = \log(y+\sqrt{(y^2+1)})$.

223. IDENTITIES FOR HYPERBOLIC FUNCTIONS

The relations $\sin iz = i \sinh z$ and $\cos iz = \cosh z$ noted in §205(v) can be used to derive hyperbolic identities from the corresponding trigonometric ones: to illustrate, put $z = iw$ in the identity $\cos 2z = 1 - 2\sin^2 z$. So

$$\cos 2iw = 1 - 2\sin^2 iw,$$

i.e. $\cosh 2w = 1 - 2(i \sinh w)^2 = 1 + 2\sinh^2 w$.

Tackle the following example with this method.

EXAMPLE. Use the corresponding trigonometric identities to show that

$$\sinh(a+b) = \sinh a \cosh b + \cosh a \sinh b,$$
$$\cosh(a+b) = \cosh a \cosh b + \sinh a \sinh b.$$

Deduce an identity for $\tanh(a+b)$. [Notice how the product of two <u>sines</u> in a trigonometric identity gives rise to a sign change in the corresponding hyperbolic identity.]

SOME OTHER RESULTS

224. DIFFERENT FUNCTIONS WITH THE SAME MACLAURIN SERIES

(i) <u>EXAMPLE</u>. The function g is defined on R by $g(x) = 1 + e^{-1/x^2}$ for $x \neq 0$ and $g(0) = 1$. Show by induction that, for every $n \epsilon N$ and for all $x \neq 0$,

$$g^{(n)}(x) = P_{3n}\left(\frac{1}{x}\right) e^{-1/x^2},$$

where P_{3n} is a polynomial of degree $3n$. Deduce that, for all $n \epsilon N$, $g^{(n)}(0)$ exists and $g^{(n)}(0) = 0$.

Show that g has $1 + 0 + 0 + 0 + 0 + \ldots$ as its Maclaurin series, (i.e. the <u>same</u> Maclaurin series as the constant function 1).

<u>Solution</u>. For $x \neq 0$, $g'(x) = \frac{2}{x^3} e^{-1/x^2} = P_3\left(\frac{1}{x}\right) e^{-1/x^2}$, as required. It is then fairly easy to show by induction

$$g^{(n)}(x) = P_{3n}\left(\frac{1}{x}\right) e^{-1/x^2} \quad (n \epsilon N, \quad x \neq 0).$$

Also, $g'(0) = \lim_{x \to 0} \dfrac{g(x) - g(0)}{x} = \lim_{x \to 0} \dfrac{1}{x} e^{-1/x^2} = \lim_{y \to \pm \infty} y e^{-y^2}$

$$= 0 \quad \text{(by the hierarchy result of §209)}.$$

Now assume that $g^{(n)}(0) = 0$ for some $n \epsilon N$. Then

$$g^{(n+1)}(0) = \lim_{x \to 0} \frac{g^{(n)}(x) - g^{(n)}(0)}{x} = \lim_{x \to 0} \frac{1}{x} P_{3n}\left(\frac{1}{x}\right) e^{-1/x^2}$$

$$= \lim_{y \to \pm \infty} y P_{3n}(y) e^{-y^2} = 0 \quad \text{(by §209)}.$$

It follows by induction that $g^{(n)}(0) = 0$ for all $n \epsilon N$.

So the Maclaurin series of g is

$$1 + 0.x + 0.\frac{x^2}{2!} + 0.\frac{x^3}{3!} + \ldots, \quad \text{i.e.} \quad 1 + 0 + 0 + 0 + \ldots.$$

N.B. This means that g has the same Maclaurin series

as the constant function 1.   The Maclaurin series, such
as it is, clearly converges to 1 and <u>not</u> to g(x) at
every non-zero point x.

(ii) <u>EXAMPLE</u>.   Give an example of a function that
is not the exponential function but which has the same
Maclaurin expansion as the exponential function.

<u>Solution</u>.   Take the function h given by $h(x) = g(x)e^x$,
where g is the function defined in part (i).   Then,
since $g^{(n)}(0) = 0$, use Leibniz's theorem (§177) to show
that $h^{(n)}(0) = 1$ for all nεN.   So h has the
Maclaurin series $1 + x + \frac{x^2}{2!} + \frac{x^3}{3!} + \dots$.

(iii) Parts (i) and (ii) show that in general
several functions can share the same Maclaurin series.
Only one of these functions however can be the <u>sum</u>
<u>function</u> of the series.

225.   <u>EXAMPLE</u>.   Prove that e is irrational.

<u>Solution</u>.   Notice that $\sum\limits_{k=0}^{\infty} \frac{1}{k!} = e$.   So,

$$0 < \left| e - (1 + 1 + \frac{1}{2!} + \frac{1}{3!} + \dots + \frac{1}{n!}) \right| = \sum_{k=n+1}^{\infty} \frac{1}{k!}$$

$$< \frac{1}{(n+1)!}\left(1 + \frac{1}{(n+1)} + \frac{1}{(n+1)^2} + \dots\right).$$

On summation of a geometric series we conclude that, for
every nεN,

$$0 < \left| e - (1 + 1 + \frac{1}{2!} + \frac{1}{3!} + \dots + \frac{1}{n!}) \right| < \frac{1}{n.n!} . \quad \dots(*)$$

Now suppose that e is in fact <u>rational</u>, so that
$e = \frac{m}{n}$ with m,nεN.   Then multiply through in (*) by n!.
The term in the middle becomes an <u>integer</u> k say.   So

$$0 < k < \frac{1}{n}.$$

This is a contradiction.   So e must be irrational.

226.   COMPLEX NUMBERS OF MODULUS ONE

Notice from §204 that $e^{i\alpha} = \cos\alpha + i\sin\alpha$, for
all αεC.   In particular notice that when α is <u>real</u>
the number $e^{i\alpha}$ has modulus equal to 1 because

$$|e^{i\alpha}| = |(\cos \alpha + i \sin \alpha)| = (\cos^2\alpha + \sin^2\alpha)^{\frac{1}{2}} = 1.$$

This means that the points $e^{i\alpha}$ ($\alpha\epsilon R$) all lie on the unit circle in the complex plane. Furthermore from the continuous and periodic behaviour of sine and cosine it is clear that <u>every</u> point z with $|z| = 1$ can be expressed as $z = \cos \alpha + i \sin \alpha = e^{i\alpha}$ for some $\alpha\epsilon[0,2\pi[$.

Notice also that $e^{2n\pi i} = 1$ for all $n\epsilon Z$.

227. THE VALUES OF THE TRIGONOMETRIC FUNCTIONS AS RATIOS OF SIDES OF A RIGHT-ANGLED TRIANGLE

Remember that we defined the real sine and cosine functions as the sum functions of power series. Their periodic behaviour, the number π and the fact that $\cos^2 x + \sin^2 x = 1$ all emerged from the definitions. Notice from these properties that the point P with coordinates $(\cos \alpha, \sin \alpha)$ where $\alpha\epsilon]0,\frac{1}{2}\pi[$ lies in the first quadrant in the xy-plane and that OP has length equal to 1. It then follows that

$$\frac{OA}{OP} = \cos \alpha,$$

$$\frac{AP}{OP} = \sin \alpha,$$

$$\frac{AP}{OA} = \tan \alpha,$$

and that the number α provides a measure of the angle AOP such that $\alpha = \frac{1}{2}\pi$ when angle $AOP = 90^\circ$.

228. COMPLEX POWERS OF POSITIVE REAL NUMBERS

By analogy with the definition of §210, we define $a^z = \exp(z \log a)$ for $a > 0$ and $z\epsilon C$. We do not pursue this idea but notice the failure of the second index law in the following example:

$$(e^{2\pi i})^i = (\cos 2\pi + i \sin 2\pi)^i = 1^i = \exp(i \log 1) = 1.$$

So, $(e^{2\pi i})^i \neq e^{2\pi i^2} = e^{-2\pi}.$

EXAMPLES 10

(It is not suggested that these be done in the order
1,2,3,4,... . It would probably be best to jump around.
Numbers 32-35,37,38,41-50 might be best left till last.)

1. Show that, for $\alpha > 0$, the graph of $x^{\alpha}\log x$ has
a minimum at the point $(e^{-1/\alpha}, -(\alpha e)^{-1})$.

2. Use the hierarchy result of §209 to show that
there exists a positive number X such that $n/\log n > 1$
for all $n > X$. Deduce that $\sum \frac{1}{\log n}$ is divergent.
Give an example of a divergent series $\sum a_n$ for which
$a_n \to 0$ as $n \to \infty$, but such that $\sum a_n^k$ is divergent for
all $k\varepsilon N$.

3. Find (a) $\lim\limits_{x\to 0} \dfrac{\cos ax - \cos bx}{x \sin x}$, (b) $\lim\limits_{x\to 1} \dfrac{x - 1}{\log x}$,

(c) $\lim\limits_{x\to\infty} e^{1/x} \cdot \dfrac{\log x}{x}$. [Put $x = 1+y$ in (b), $x = \frac{1}{y}$ in (c).]

4. Use the mean-value theorem to prove that
$$\frac{1}{n + 1} < \log(n + 1) - \log n < \frac{1}{n} \qquad (n\varepsilon N).$$
Hence discuss the convergence of $\sum \left(\log(1 + \frac{1}{n})\right)^k$ for all
values of $k > 0$.

5. Use §213 to show that, for all $n\varepsilon N$,
$$\log(n + 1) = \log n + 2\left[\frac{1}{(2n + 1)} + \frac{1}{3(2n + 1)^3} + \cdots\right].$$

6. Discuss the behaviour as $n \to \infty$ of

(a) $\left(\dfrac{n}{n - 1}\right)^n$, (b) $\left(\dfrac{n^2 - 1}{n^2 + 1}\right)^n$, (c) $n(2^{1/n^2} - 1)$, (d) $n^{1/\log n}$,

(e) $n(a^{1/n} - b^{1/n})$, (f) $\sqrt{n}(n^{1/n} - 1)$, (g) $5^{n^2}n^{-n}$.

7. Apply the mean-value theorem to $\log x$ on $[1,a]$,
where $a > 1$, to show that
$$1 - \frac{1}{a} < \log a < a - 1.$$
Show that this inequality also holds for $a\varepsilon]0,1[$.

8.   Find $\displaystyle\sum_{n=1}^{\infty} i^n/n$ explicitly.

9.   Discuss the convergence of

(a) $\sum \dfrac{\log n}{n}$,   (b) $\sum \dfrac{1}{n \log n}$,

(c) $\sum \dfrac{1}{n(\log n)^2}$,   (d) $\sum \dfrac{1}{(\log n)^n}$.

10.   (<u>Logarithms to the base 10</u>) Notice that we sometimes denote the logarithmic function defined in §207 by $\log_e x$ instead of just $\log x$ and we can talk of it as the <u>logarithm to the base e</u> to emphasise that it is as the inverse of e^x that it is derived. Logarithms to other bases arise in the following way.

Define $f: R \to \,]0,\infty[$ by $f(x) = 10^x$.   Show that f is bijective.   The value of its inverse at y $(y > 0)$ is denoted by $\log_{10} y$.   Show that $\dfrac{d}{dy}(\log_{10} y) = \dfrac{1}{y \log_e 10}$.
[Here $\log_{10} y$ is the <u>logarithm of y to the base 10</u>.]

11.   Show that for positive numbers a,b,c,

$$\log_a b \cdot \log_b c = \log_a c .$$

Deduce that $\log_a b = 1/\log_b a$.

12.   It is given that $\log_{10} e = 0.4342945$ (7 places). Use the mean-value theorem to show that

$$\frac{3}{128} \log_{10} e < \log_{10} 128 - \log_{10} 125 < \frac{3}{125} \log_{10} e.$$

Deduce that $\log_{10} 2$ lies between 0.301018 and 0.301042. (You will need the results of Exs.10 & 11 above.)

13.   Use a calculator to evaluate successively $3^{10}, \; 3^{100}, \; \ldots$ and show that 3^{100000} lies between 10^{47712} and 10^{47713}.   Deduce that $\log_{10} 3 = 0.4771$ (4 places).

14.   State the Maclaurin series of 2^x.

15.   Show that $\sin^{-1} x$ satisfies the differential equation $(1 - x^2)y'' - xy' = 0$ and hence show that the Maclaurin series of $\sin^{-1} x$ is

$$x + \frac{1^2}{3!}\,x^3 + \frac{1^2 \cdot 3^2}{5!}\,x^5 + \frac{1^2 \cdot 3^2 \cdot 5^2}{7!}\,x^7 + \ldots \;.$$

16. Use the idea of Ex.7.14 to show that, for $x \varepsilon [-1,1[$,

$$\log(1 + x + x^2) = x + \frac{x^2}{2} - \frac{2x^3}{3} + \frac{x^4}{4} + \frac{x^5}{5} - \frac{2x^6}{6} + \ldots \;.$$

17. Show that, for $x \varepsilon [0,1]$,

$$x - \tfrac{1}{2}x^2 \le \log(1 + x) \le x \;.$$

18. By doing estimation on the tail of the series for $\log(1 + x)$, prove that

(a) $\left|\log(1 + x)\right| \le 2|x|$ for $|x| \le \tfrac{1}{2}$,

(b) $\left|\log(1 + x) - x\right| \le x^2$ for $|x| \le \tfrac{1}{2}$.

19. Prove that, for all $x \varepsilon R$,

$$\sin 3x = 3\sin x - 4\sin^3 x, \qquad \tan 3x = \frac{3\tan x - \tan^3 x}{1 - 3\tan^2 x}\;.$$
$$\cos 3x = 4\cos^3 x - 3\cos x,$$

20. Find the identities for hyperbolic functions corresponding to

$$\cos 2z = \frac{1 - \tan^2 z}{1 + \tan^2 z}\,, \qquad \sin 2z = \frac{2\tan z}{1 + \tan^2 z}\,, \qquad \sec^2 z = 1 + \tan^2 z.$$

21. Show that, for all $n \varepsilon N$,

$$\tan^{-1}\frac{1}{n} - \tan^{-1}\frac{1}{n+1} = \tan^{-1}\left(\frac{1}{n^2+n+1}\right).$$

Deduce that $\displaystyle\sum_{n=1}^{\infty} \tan^{-1}\left(\frac{1}{n^2+n+1}\right) = \frac{\pi}{4}$.

22. Show that, for $x \varepsilon\,]-1,1[$,

$$2\tan^{-1}x - \tan^{-1}2x = \tan^{-1}\left(\frac{2x^3}{1 + 3x^2}\right).$$

23. Notice that (to the accuracy given)

$\log_{10}e = 0.43, \quad \log_{10}1.01 = 0.0043, \quad \log_{10}1.001 = 0.00043.$
Explain.

24. Use the hierarchy result (§209) to show that there exist numbers M and N such that

$$x^{10}e^{-x} < 1/x^2 \quad \text{for all} \quad x > M$$

and $\displaystyle\frac{1}{(\log x)^{10}\sqrt{(x + 1)}} \ge \frac{1}{x^{3/4}}$ for all $x > N.$

25.   Find the radius of convergence R of $\sum \frac{n^{2n}}{(2n)!} z^n$.
Use Stirling's formula from §270 to settle
whether the series converges for $z = R$.

26.   Discuss the behaviour as $n \to \infty$ of

(a) $\left[1 + \frac{\log n}{n}\right]^n$,   (b) $\left[1 + \frac{1}{n \log n}\right]^n$,   (c) $\left[1 + \frac{1}{\sqrt{n}}\right]^n$,

(d) $\left|\left(1 + \frac{i}{n}\right)^n\right|$,   (e) $n\left[\left(1 + \frac{1}{n^2}\right)^n - 1\right]$,

(f) $\left[\frac{\log(n+1)}{\log n}\right]^n$,   (g) $(n!)^{1/n}$.

(For (g) use Ex.8.3 or Stirling's formula in §270.)

27.   Find the limits as $x \to 0+$ of (a) x^x, (b) x^{x^x}.
Find $(.01)^{.01}$ and $.01^{.01^{.01}}$ on a calculator and compare.

28.   (To solve $e^x = \tan x$) By considering the
behaviour as $x \to 0$ and as $x \to \frac{1}{2}\pi$ of $(e^x - \tan x)$,
prove that the equation $e^x = \tan x$ has at least one
root in $]0, \frac{1}{2}\pi[$.
    By writing the equation as $x = \tan^{-1}(e^x)$ and using
the mean-value theorem, prove that there is exactly one
such root α. (See §135(i).)
    Define a sequence by taking a_1 arbitrarily in
$]0, \frac{1}{2}\pi[$ and $a_{n+1} = \tan^{-1}(e^{a_n})$ for $n\epsilon N$.   Prove that
$|a_{n+1} - \alpha| < \frac{1}{2}|a_n - \alpha|$ and deduce that $a_n \to \alpha$ as $n \to \infty$.
[Use a calculator to show that $\alpha = 1.306$ (3 places).]

29.   (To solve $x \tan x = 2$)   Prove that the equation
$x \tan x = 2$ has exactly one root in $]0, \frac{1}{2}\pi[$.
    Define a sequence $\{a_n\}$ by taking a_1 arbitrarily
in $]0, \frac{1}{2}\pi[$ and $a_{n+1} = \tan^{-1}(2/a_n)$ for $n\epsilon N$.   Use the
mean-value theorem to estimate $|a_{n+1} - \alpha|$ and deduce
that $a_n \to \alpha$ as $n \to \infty$.   [Use a calculator to show that
$\alpha = 1.077$ (3 places).]

30. (To solve $e^{3x} = \tan x$) A sequence is defined by taking $a_1 \in R$ and $a_{n+1} = \tan^{-1}(e^{3a_n})$ for all $n \in N$. Prove that the sequence $\{a_n\}$ is monotonic and bounded and deduce that $\{a_n\}$ tends to a limit α as $n \to \infty$. [Use a calculator to show that $\alpha = 1.5616$ (4 places).]

31. Use the power series for $\cos x$ and $\log(1 + x)$ to show that $\log(\cos x) = -\dfrac{x^2}{2} - \dfrac{x^4}{12} - \dfrac{x^6}{45} + O(x^8)$ for $x \in]0, r[$, where r is a positive number. Deduce that the Maclaurin series of $\tan x$ is $x + \dfrac{x^3}{3} + \dfrac{2x^5}{15} + O(x^7)$. (Actually the radii of convergence of both these series is $\frac{1}{2}\pi$.)

32. Show that $\displaystyle\sum_{n=1}^{\infty} \dfrac{(n + 1)^2}{n!} = 5e - 1$.

33. Let $z = x + iy$. Prove that $\left| \left(1 + \dfrac{z}{n}\right)^n \right| \to e^x$ as $n \to \infty$.

34. Prove that as $x \to \infty$, $e^{(\log x)^2}$ tends to infinity slower than e^x but faster than every positive power of x, (i.e. $e^{(\log x)^2}$ lies between e^x and x^k in the hierarchy of §209). (See also the solution of Ex.13.18.)

35. By noticing that $\log(2n) - \log n = 0.6931\ldots$, show it is possible to choose arbitrarily large positive integers n such that $\sin(\log(n+1))$, $\sin(\log(n+2))$, ..., $\sin(\log(2n))$ are all greater than $1/\sqrt{2}$. Deduce that $\displaystyle\sum \dfrac{\sin(\log n)}{n}$ diverges.

36. Starting from the fact that $\sin \frac{1}{2}\pi = 1$, use identities to show that $\sin \dfrac{\pi}{4} = \cos \dfrac{\pi}{4} = 1/\sqrt{2}$.

37. Is $\log x$ uniformly continuous on (a) $]0,1]$, (b) $]1,\infty[$?

38. Prove that $\cosh^2 z = \frac{1}{2}(1 + \cosh 2z)$ by multiplication of series.

39.　By considering its derivative prove that
$\tan^{-1}y + \tan^{-1}(1/y)$ is constant for all $y > 0$. What happens for $y < 0$?

40.　Show that $x^{1/x}$ attains its maximum value for $x > 0$ at $x = e$.

41.　Show that, for $|a| < 1$ and $\alpha \in R$,
$$\sum_{n=1}^{\infty} a^n \sin n\alpha = \frac{a \sin \alpha}{1 - 2a \cos\alpha + a^2} \, .$$
Deduce that, for $|a| > 1$,
$$\sum_{n=1}^{\infty} a^{-n} \sin n\alpha = \frac{a \sin \alpha}{1 - 2a \cos\alpha + a^2} \, .$$

42.　By splitting the series
$$\frac{x}{1.2} - \frac{x^2}{3.4} + \frac{x^3}{5.6} - \dots \qquad (-1 < x < 1)$$
into two parts, find an explicit expression for its sum.

43.　Find an explicit expression for the sum of
$$x + \frac{x^5}{5} + \frac{x^9}{9} + \dots \qquad (-1 < x < 1).$$

44.　What is the period of the exponential function?

45.　Suppose that $0 < a < 1$. Find the limits as $n \to \infty$ of (a) $(1 + a^n)^n$, (b) $(1 + a^n)^{n!}$.

46.　Show that $\left| \left(1 + \frac{i}{\sqrt{n}} \right)^n \right|$ tends to a limit as $n \to \infty$.

47.　Show that, for $x > 1$, $(\log x)^{\log x} = x^{\log\log x}$. Deduce that $\exists\, N$ such that $(\log x)^{\log x} > x^2 \ \forall x > N$.

48.　In Ex.34 above, the function $e^{(\log x)^2}$ is found to lie between x^k and e^x in the hierarchy of §209. Find a similar function lying between $\log x$ and x^k.

49.　Find the limit as $n \to \infty$ of $\left(\frac{1 + a^{1/n}}{2} \right)^n$.

50.　Find (a) $\lim_{x \to 0+} e^{(\log x)^2}$, (b) $\lim_{x \to 0+} e^{(\log x)^3}$,

(c) $\lim\limits_{x \to 0+} (1 + \log(1 + x))^{1/x}$.

51.   By considering its derivative show that the function $(\tan x - x)$ is increasing on $]0, \tfrac{1}{2}\pi[$.   Deduce that $\tan x > x$ for all $x \in]0, \tfrac{1}{2}\pi[$.

52.   Show that $x > \tan^{-1} x > x - \dfrac{x^3}{3}$ for all $x > 0$.

53.   Differentiation of rational powers is covered in §124: for rational α and for suitable values of y the result is that

$$\frac{d}{dy}(y^\alpha) = \alpha y^{\alpha - 1}.$$

Irrational powers are defined in §210, the definition being that $y^\alpha = e^{\alpha \log y}$.   Use this definition to show that, for irrational α and for suitable values of y,

$$\frac{d}{dy}(y^\alpha) = \alpha y^{\alpha - 1}.$$

XI. INTEGRATION

---

229. Integration not only permits evaluation of the area under a curve but also provides a method for evaluating certain limits and for developing power series expansions.

There are two aspects to consider - <u>indefinite integration</u> and <u>Riemann integration</u>. The former is defined as the reverse of differentiation while the latter is defined as the finding of the area of a region under a curve. The connection between the two ideas is made in §245.

INDEFINITE INTEGRATION

230. For a real function f defined on an interval I in R, any function F such that $F'(x) = f(x)$ for all $x \varepsilon I$ (with suitable conventions at endpoints) is called an <u>indefinite integral</u> of f on I. For example, x^3 and $x^3 + 8$ are both indefinite integrals of $3x^2$ on R.

Actually if two functions F_1 and F_2 are both indefinite integrals of f on [a,b], then they <u>must</u> differ by a constant. For the derivative of $F_1 - F_2$ is zero on [a,b], and as noted in §130, this is enough to force $F_1 - F_2$ to be a constant function.

Methods of finding indefinite integrals and in particular the methods of integration by parts, by change of variable and by partial fractions are treated in books on Calculus and are not discussed here. The following list is worth knowing.

function f	indefinite integral F
$x^k \quad (k \neq -1)$	$\dfrac{x^{k+1}}{k+1}$
$1/x$	$\log |x|$
e^x	e^x
$\log x$	$x \log x - x$
$\sin x$	$- \cos x$
$\cos x$	$\sin x$
$\tan x$	$\log|\sec x|$
$\operatorname{cosec} x$	$- \log |\operatorname{cosec} x + \cot x|$
$\sec x$	$\log|\sec x + \tan x|$

function f	indefinite integral F
$\cot x$	$\log|\sin x|$
$\dfrac{1}{x^2 + a^2}$	$\dfrac{1}{a} \tan^{-1} \dfrac{x}{a}$
$\dfrac{1}{\sqrt{(x^2 + a^2)}}$	$\log(x + \sqrt{(x^2 + a^2)})$
$\dfrac{1}{\sqrt{(x^2 - a^2)}}$	$\log|x + \sqrt{(x^2 - a^2)}|$
$\dfrac{1}{\sqrt{(a^2 - x^2)}}$	$\sin^{-1} \dfrac{x}{a}$

RIEMANN INTEGRATION

231. A <u>dissection</u> of a closed interval [a,b] is a <u>finite</u> set $D = \{c_0, c_1, \ldots, c_n\}$, where
$$a = c_0 < c_1 < c_2 < \ldots < c_n = b.$$

<u>Refining</u> the dissection D (i.e. adding more points to D), produces a <u>finer</u> dissection.

232. RIEMANN SUMS

Let f be a bounded real function on [a,b].

(i) Let A be the region enclosed by the lines $x = a$, $x = b$, the x-axis and the curve $y = f(x)$.

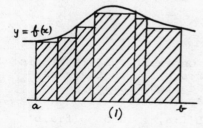 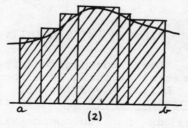

(1) $\qquad\qquad\qquad$ (2)

In general, corresponding to dissection D of [a,b] with points $c_0, \ldots, c_n$, we can define the <u>lower</u> and <u>upper Riemann sums</u> by

$$L(f:D) = \sum_{k=1}^{n} m_k(c_k - c_{k-1}) \quad \text{and} \quad U(f:D) = \sum_{k=1}^{n} M_k(c_k - c_{k-1}),$$

where m_k and M_k denote the infimum and supremum of f on $[c_{k-1}, c_k]$ respectively. These sums measure the areas of the rectangles shown in the diagram (1) and (2) respectively, and thereby provide lower and upper bounds for the area of the region A.

Notice that in the diagrams f is taken to be non-negative and continuous. Neither of these features is essential. Notice also that, for every dissection D,

$$L(f:D) \leq \text{area of } A \leq U(f:D) .$$

(ii) Notice that L(f:D) <u>increases</u> and U(f:D) <u>decreases</u> as D is refined. To check this (<u>either</u> by looking at the areas of rectangles <u>or</u> analytically) consider the effect on the upper and lower sums of inserting <u>one</u> new point of dissection between two points of the original dissection D.

(iii) Suppose that m and M denote inf f and sup f on [a,b]. Then, by (ii), for every dissection D of [a,b],

$$m(b-a) \leq L(f:D) \leq U(f:D) \leq M(b-a). \qquad \dots(*)$$

It follows that, for every bounded function f, the numbers

$$\sup_{D} L(f:D) \quad \text{and} \quad \inf_{D} U(f:D)$$

exist. (The notation means that the supremum and infimum are taken over all the values obtained as D runs through all possible dissections of [a,b].)

Denote

$$\sup_{D} L(f:D) \text{ by } L(f:a,b) \quad \text{and} \quad \inf_{D} U(f:D) \text{ by } U(f:a,b) .$$

It then follows from (*) that

$$L(f:a,b) \leq U(f:a,b) .$$

233. RIEMANN INTEGRABILITY

For a bounded function f on [a,b], L(f:a,b) is in general strictly less than U(f:a,b). In the case when L(f:a,b) = U(f:a,b), f is called <u>Riemann integrable on [a,b]</u>. In this case we write $\int_{a}^{b} f(x)\, dx$ for the common value L(f:a,b) and U(f:a,b) and call it the <u>value</u> of the integral of f on [a,b].

From §232, $\int_{a}^{b} f(x)\, dx$ clearly measures the signed area enclosed by the x-axis, the curve y = f(x) and the lines x = a and x = b.

Let B[a,b], C[a,b], M[a,b] and R[a,b] denote the sets of real functions on [a,b] that are bounded, continuous, monotonic and Riemann integrable respectively. Then the following inclusions are worth noting:

$R[a,b] \subsetneq B[a,b]$; $C[a,b] \subsetneq R[a,b]$; $M[a,b] \subsetneq R[a,b]$.

These three inclusions are all strict.   See §234, §238 and §240 for more about them.

234.   A BOUNDED FUNCTION THAT IS NOT RIEMANN INTEGRABLE

EXAMPLE.   Let f be defined on [0,6] by

$f(x) = 0$ (x rational) and $f(x) = 1$ (x irrational).

Show that f is not Riemann integrable on [0,6].

Solution.   Let D be a dissection of [0,6] with points $c_0, \ldots, c_n$.   Then, since there are both rational and irrational points in each interval $[c_{k-1}, c_k]$, it follows that, for every k, we have $m_k = 0$ and $M_k = 1$.   So, for every dissection D,

$$L(f:D) = \sum_{k=1}^{n} m_k(c_k - c_{k-1}) = \sum_{k=1}^{n} 0(c_k - c_{k-1}) = 0,$$

$$U(f:D) = \sum_{k=1}^{n} M_k(c_k - c_{k-1}) = \sum_{k=1}^{n} 1(c_k - c_{k-1}) = 6.$$

So $L(f:0,6) = 0$ while $U(f:0,6) = 6$.   So $f \notin R[0,6]$.

235.   A dissection function is a function which takes
a value for each dissection of [a,b].   Examples are $L(f:D)$ and $U(f:D)$ in §232.

That the dissection function H tends to the real number K under refinement means:

$\forall \varepsilon > 0$, $\exists$ a dissection D_1 such that

$$|H(D) - K| < \varepsilon \quad \text{for all D finer than } D_1.$$

(As notation for this take $\lim_D H(D) = K$.)

Notice that in fact $L(f:a,b) = \lim_D L(f:D)$.   For since $L(f:a,b) = \sup_D L(f:D)$, there exists a dissection D_1 such that $L(f:a,b) - \varepsilon < L(f:D_1) \leq L(f:a,b)$. Then, since $L(f:D)$ increases under refinement (§232), for every D finer than D_1 we have $L(f:D_1) \leq L(f:D)$. So, for every D finer than D_1, we have that $L(f:D)$ is within ε of $L(f:a,b)$, i.e. $L(f:a,b) = \lim_D L(f:D)$.

Similarly, $U(f:a,b) = \lim_D U(f:D)$.

G

236. In the situation of §232 notice the dissection functions Δ and S defined by:

$$\Delta(f:D) = \sum_{k=1}^{n} (M_k-m_k)(c_k-c_{k-1})$$

and

$$S(f:D,e) = \sum_{k=1}^{n} f(e_k)(c_k-c_{k-1}),$$

where each e_k is an arbitrarily chosen point in the interval $[c_{k-1},c_k]$.

Notice that $\Delta(f:D)$ <u>decreases</u> as D is refined: this is a consequence of the behaviour of U and $-L$ mentioned in §232. It follows from §233 and §235 that, for a bounded function f,

$$f \in R[a,b] \quad \text{if and only if} \quad \lim_{D} \Delta(f:D) = 0.$$

It also follows that

$$L(f:D) \leq S(f:D,e) \leq U(f:D),$$

because $m_k \leq f(e_k) \leq M_k$ whenever $e_k \in [c_{k-1},c_k]$. So it follows by a sandwiching argument that if $f \in R[a,b]$ then $\lim_{D} S(f:D,e) = \int_a^b f(x)\,dx$ for all choices of the points e_k.

It is also true that if $\lim_{D} S(f:D,e) = K$ for all choices of the points e_k, then $f \in R[a,b]$ and $\int_a^b f(x)\,dx = K$.

237. HEREDITY RESULTS FOR RIEMANN INTEGRABILITY

Notice first that for a general bounded function f on a subinterval $[c_{k-1},c_k]$ there need <u>not</u> exist points at which m_k and M_k are attained. (Remember that m_k and M_k are the infimum and supremum of f on $[c_{k-1},c_k]$.) However, for every $\varepsilon > 0$, there do exist points p_k and q_k in $[c_{k-1},c_k]$ such that

$$f(p_k) < m_k + \varepsilon \quad \text{and} \quad f(q_k) > M_k - \varepsilon$$

so that

$$M_k - m_k < (f(q_k) - f(p_k)) + 2\varepsilon. \qquad \dots(*)$$

This is useful in the proof of the following result.

<u>RESULT.</u> Let f and g be bounded Riemann integrable functions on $[a,b]$ and let $k \in R$. Then each of the following functions is also Riemann integrable on $[a,b]$:

$$f+g, \quad kf, \quad fg, \quad |f|, \quad \max(f,g), \quad \min(f,g).$$

Proof. Since $f, g \in R[a,b]$, notice that
$$\lim_D \Delta(f:D) = \lim_D \Delta(g:D) = 0.$$

For a dissection D of $[a,b]$ with points c_i, use the equivalent of (*) on page 186 for the function $f + g$ to conclude the existence of points q_k, p_k for $k = 1, 2, \ldots, n$ such that

$$\Delta(f+g:D) \leq \sum_{k=1}^{n} [(f+g)(q_k) - (f+g)(p_k)](c_k - c_{k-1})$$

$$+ \sum_{k=1}^{n} 2\varepsilon(c_k - c_{k-1})$$

$$= \sum_{k=1}^{n} (f(q_k) - f(p_k))(c_k - c_{k-1})$$

$$+ \sum_{k=1}^{n} (g(q_k) - g(p_k))(c_k - c_{k-1}) + 2\varepsilon(b-a)$$

$$\leq \Delta(f:D) + \Delta(g:D) + 2\varepsilon(b-a)$$

$$< K\varepsilon \quad \text{for all } D \text{ finer than } D_1, \text{ say,}$$

where K is a constant.
So $\lim_D \Delta(f+g:D) = 0.$ So $(f+g) \in R[a,b]$.

For kf use the fact that , for $k > 0$, $U(kf:D) = k\,U(f:D)$ etc. In particular, $-f \in R[a,b]$ and hence $(f-g) \in R[a,b]$.

Then, to show that $fg \in R[a,b]$ it will be enough to show that $f^2 \in R[a,b]$, because $4fg = (f+g)^2 - (f-g)^2$. So, apply (*) on page 186 to f^2 to conclude the existence of points q_k, p_k $(k = 1, \ldots, n)$ such that

$$\Delta(f^2:D) \leq \sum_{k=1}^{n} [(f(q_k))^2 - (f(p_k))^2](c_k - c_{k-1}) + 2\varepsilon(b-a)$$

$$\leq \sum_{k=1}^{n} (f(q_k) - f(p_k))(f(q_k) + f(p_k))(c_k - c_{k-1}) + 2\varepsilon(b-a)$$

$$\leq 2M \sum_{k=1}^{n} (f(q_k) - f(p_k))(c_k - c_{k-1}) + 2\varepsilon(b-a)$$

$$\leq 2M\,\Delta(f:D) + 2\varepsilon(b-a)$$

$$< K'\varepsilon \quad \text{for all } D \text{ finer than } D_2, \text{ say, where}$$

K' is a constant.
So $\lim_D \Delta(f^2:D) = 0.$ So $f^2 \in R[a,b]$.

188

Finally in the proof of the heredity result, use the fact that $||a| - |b|| \leq |a-b|$ together with (*) on page 186 to see that $|f| \varepsilon R[a,b]$. Couple this with the formulae of §5 to show that $\max(f,g)$ and $\min(f,g)$ are both in $R[a,b]$ too.

N.B. Using $S(f:D,e)$ it is easy to show that

$$\int_a^b (f+g)(x) \ dx = \int_a^b f(x) \ dx + \int_a^b g(x) \ dx$$

and $\qquad \int_a^b (kf)(x) \ dx = k\int_a^b f(x) \ dx$, for a constant k.

In the language of Functional Analysis this means that $\int_a^b . \ dx$ is a <u>linear functional</u> on $R[a,b]$.

238. A CONTINUOUS FUNCTION ON A CLOSED INTERVAL IS RIEMANN INTEGRABLE

Call the maximum length of the subintervals of a dissection D its <u>mesh length</u>. Denote the mesh length by $|D|$.

<u>RESULT</u>. $C[a,b] \subsetneq R[a,b]$.

<u>Proof</u>. Let $f \varepsilon C[a,b]$. Then f is uniformly continuous on $[a,b]$ by §111. So
$$\forall \varepsilon > 0, \quad \exists \delta > 0 \quad \text{such that} \quad |f(x) - f(y)| < \varepsilon/(b-a)$$
$$\forall x,y \quad \text{with} \quad |x-y| < \delta.$$
Now take a dissection D_1 of $[a,b]$ with $|D| < \delta$.

Then $\quad \Delta(f:D_1) = \sum_{k=1}^{n} (M_k - m_k)(c_k - c_{k-1})$

$$\leq \sum_{k=1}^{n} \frac{\varepsilon}{(b-a)} . (c_k - c_{k-1}) = \frac{\varepsilon}{(b-a)} . (b-a) = \varepsilon.$$

Then, since Δ decreases as D is refined, see that $\Delta(f:D) < \varepsilon$ for all D finer than D_1. So $\lim_D \Delta(f:D) = 0$ and we conclude that $f \varepsilon R[a,b]$.

239. DISSECTIONS WITH MESH LENGTH TENDING TO ZERO

<u>RESULT</u>. Let $f \varepsilon C[a,b]$ and let $\{D_n\}$ be a sequence of dissections of $[a,b]$ such that $|D_n| \to 0$ as $n \to \infty$. Then $L(f:D_n)$ and $U(f:D_n)$ tend to $\int_a^b f(x) \ dx$ as $n \to \infty$.

Proof.   Use uniform continuity as in §238 to see that $\Delta(f:D_n) \to 0$ as $|D_n| \to 0$.   (A prescribed ε dictates a δ and we then choose n sufficiently large to make $|D_n| < \delta$.)   So $U(f:D_n) - L(f:D_n) \to 0$ as $n \to \infty$. But $L(f:D_n) \le \int_a^b f(x)\, dx \le U(f:D_n)$ for every n. Hence the result.

The value of this result in the evaluation of limits can be seen in §246.

240.   A MONOTONIC FUNCTION ON A CLOSED INTERVAL IS RIEMANN INTEGRABLE

RESULT. $M[a,b] \subseteq R[a,b]$

Proof.   Choose $\varepsilon > 0$.   Suppose without loss of generality that f is increasing and non-constant on [a,b].   Choose a dissection D_1 of [a,b] with mesh length less than $\varepsilon/(f(b)-f(a))$.   Then

$$\Delta(f:D_1) = \sum_{k=1}^{n} (f(c_k) - f(c_{k-1}))(c_k - c_{k-1})$$

$$\le \sum_{k=1}^{n} (f(c_k) - f(c_{k-1})) \cdot \frac{\varepsilon}{f(b)-f(a)}$$

$$= \frac{\varepsilon}{f(b)-f(a)} \cdot \sum_{k=1}^{n} (f(c_k)-f(c_{k-1}))$$

$$= \frac{\varepsilon}{f(b)-f(a)} \cdot (f(b)-f(a))$$

$$= \varepsilon.$$

So, as in §238, we see that $f \varepsilon R[a,b]$.

Two valuable classes of Riemann integrable functions are identified by the results of §238 and §240, namely continuous functions and monotonic functions.   The example of §241 provides further insight in this direction.

241. <u>EXAMPLE</u>. Let f be defined on [0,1] by

f(x) = 8 when x = ½, f(x) = 0 otherwise.

Prove that f ∈ R[0,1] and that $\int_0^1 f(x)\,dx = 0$.

<u>Solution</u>. Choose ε > 0 and without loss of generality let ε < ½. Then let D_1 be the dissection of [0,1] with points 0, ½-ε, ½+ε, 1. Then

$$L(f:D_1) = 0.(½-ε) + 0.2ε + 0.(½-ε) = 0$$

and

$$U(f:D_1) = 0.(½-ε) + 8.2ε + 0.(½-ε) = 16ε.$$

Since L(f:D) increases and U(f:D) decreases as D is refined, it follows that $\lim_D L(f:D) = \lim_D U(f:D) = 0$.

So f ∈ R[0,1] and $\int_0^1 f(x)\,dx = 0$.

Notice that a similar argument will show that a function g that is zero except at a <u>finite</u> number of points of a closed interval [a,b] is Riemann integrable on [a,b] and the value of the integral is zero.

As a further extension of this last comment notice that if f and g are two Riemann integrable functions that differ at only a <u>finite</u> number of points then $\int_a^b f(x)\,dx = \int_a^b g(x)\,dx$. For f-g ∈ R[a,b] and $\int_a^b (f(x)-g(x))\,dx = 0$ by the above.

242. To write $\int_0^{½π} \frac{\sin x}{x}\,dx$ may appear dangerous

because the integrand is undefined at x = 0. Notice however that $\frac{\sin x}{x} \to 1$ as x → 0+ and the function F defined on [0,½π] by

$$F(x) = \frac{\sin x}{x} \quad (x \neq 0), \quad F(0) = 1$$

is continuous and so Riemann integrable on [0,½π]. In a case like this we take $\int_0^{½π} \frac{\sin x}{x}\,dx = \int_0^{½π} F(x)\,dx$.

243. TWO INEQUALITIES FOR INTEGRALS

The following result is simple but very useful.

<u>RESULT</u>. Let f ∈ R[a,b]. Let m and M be lower and upper bounds for f on [a,b]. Then the following inequalities hold:

$$\text{(i)} \quad m(b-a) \leq \int_a^b f(x) \, dx \leq M(b-a);$$

$$\text{(ii)} \quad \left| \int_a^b f(x) \, dx \right| \leq \int_a^b |f(x)| \, dx.$$

Proof. For (i) notice that

$$\int_a^b (M-f(x)) \, dx \geq 0 \quad \text{and} \quad \int_a^b (f(x)-m) \, dx \geq 0,$$

since both integrands are non-negative.

For (ii) notice that for similar reasons

$$\int_a^b (|f(x)|-f(x)) \, dx \geq 0 \quad \text{and} \quad \int_a^b (|f(x)|+f(x)) \, dx \geq 0.$$

Then apply the idea of §7.

244. By looking at dissections it can be shown that $f \in R[a,b]$ if and only if $f \in R[x,y]$ for every subinterval $[x,y]$ of $[a,b]$. Also notice that if $f \in R[a,b]$ and $a \leq c \leq d \leq b$ then

$$\int_a^d f(x) \, dx = \int_a^c f(x) \, dx + \int_c^d f(x) \, dx;$$

so then $\int_c^d f(x) \, dx = \int_a^d f(x) \, dx - \int_a^c f(x) \, dx.$ ...(*)

Furthermore, for $d < c$, we now define

$$\int_c^d f(x) \, dx = -\int_d^c f(x) \, dx.$$

The relation (*) then holds for all $c,d \in [a,b]$.

245. THE CONNECTION BETWEEN INDEFINITE INTEGRATION AND RIEMANN INTEGRATION

For a given function $f \in R[a,b]$, Riemann integration allows us to calculate (by dissection limits) a number, namely $\int_a^x f(t) \, dt$, for each $x \in [a,b]$. This lets us define a function $F: [a,b] \to R$ by

$$F(x) = \int_a^x f(t) \, dt. \qquad \qquad \ldots(\alpha)$$

The number $F(c)$ can then be thought of as the area of the region bounded by the curve $y = f(x)$, the lines $x = a$, $x = c$ and the x-axis.

It is important to realise that if f is continuous on $[a,b]$ then F is actually an indefinite integral of f. This is the message of the following result.

RESULT. Let $f \in R[a,b]$. Let $F: [a,b] \to R$ be defined as in (α) above. Let c be a point of $[a,b]$ at which f is continuous. Then F is differentiable at c and $F'(c) = f(c)$.

Proof.   Choose $\varepsilon > 0$.   Use the continuity of f at c
to conclude the existence of $\delta > 0$ such that

$f(t) \varepsilon \ ]f(c) - \varepsilon , f(c) + \varepsilon [$ whenever $t \varepsilon \ ]c - \delta , c + \delta [$.

Now from §244 notice that

$$F(c+h) - F(c) = \int_c^{c+h} f(t) \, dt.$$

Use the estimate (i) of §243 to see that, for all h
with $0 \leq |h| < \delta$, $(F(c+h) - F(c))$ lies between
$h(f(c) - \varepsilon)$ and $h(f(c) + \varepsilon)$.   So this means that
$\dfrac{F(c+h) - F(c)}{h}$ lies between $f(c) - \varepsilon$ and $f(c) + \varepsilon$,

whenever $|h| < \delta$,   i.e.

$$\lim_{h \to 0} \frac{F(c+h) - F(c)}{h} = f(c), \quad \text{i.e.} \quad F'(c) = f(c).$$

N.B.   Suppose that f is continuous on [a,b].   From
§244 notice that

$$\int_c^d f(t) \, dt = F(d) - F(c),$$

where F is defined by (α) on page 191.   Notice also
from the last result that F is actually an indefinite
integral of f on [a,b] and that as such it differs
from any other indefinite integral by a constant function
as mentioned in §230.   Consequently, if G is any
indefinite integral of f on [a,b] (no matter how G
is found), then

$$\int_c^d f(t) \, dt = G(d) - G(c). \qquad \ldots (\beta)$$

In general, finding an indefinite integral is easier than
evaluating a dissection limit.   (β) reduces the problem
of finding a dissection limit (or equivalently, finding
the area of a region under a curve) to the finding of an
indefinite integral.   This is exploited in §246.

APPLICATIONS OF INTEGRATION TO LIMITS AND POWER SERIES

246.   FINDING CERTAIN LIMITS USING INTEGRATION

EXAMPLE.   Find $\int_1^6 \frac{1}{x} dx$ and hence evaluate

$\lim_{n \to \infty} \left[\dfrac{1}{n+1} + \dfrac{1}{n+2} + \ldots + \dfrac{1}{6n} \right]$.

Solution.   Take $f(x) = 1/x$ for $x \varepsilon [1,6]$.   Then
because f is continuous on [1,6], $f \varepsilon R[1,6]$ by §238.

Let D_n be the dissection of $[1,6]$ with points $1, 1 + \frac{1}{n}, 1 + \frac{2}{n}, \ldots, 6 - \frac{1}{n}, 6$. Then D_n has mesh length $\frac{1}{n}$. So by §239 see that $L(f:D_n) \to \int_1^6 \frac{1}{x} dx$ as $n \to \infty$.

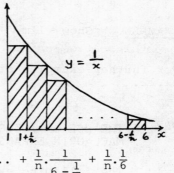

So $L(f:D_n) = \frac{1}{n} \cdot \frac{1}{1 + \frac{1}{n}} + \frac{1}{n} \cdot \frac{1}{1 + \frac{2}{n}} + \ldots + \frac{1}{n} \cdot \frac{1}{6 - \frac{1}{n}} + \frac{1}{n} \cdot \frac{1}{6}$

$$= \frac{1}{n+1} + \frac{1}{n+2} + \ldots + \frac{1}{6n-1} + \frac{1}{6n}$$

$$\to \int_1^6 \frac{1}{x} dx = [\log x]_1^6 = \log 6 - \log 1 = \log 6.$$

N.B.   By making an arbitrary choice of integrand and range of integration we can evaluate the resulting limit as in the above example.   Suppose however that we are confronted with the limit on its own. How then do we get the integrand and the range of integration?   In this situation often it helps to take a dissection with subintervals of length $1/n$.   This means making each term in the original limit into

$$\frac{1}{n} \times (\text{another factor}).$$

This corresponds to the area of a rectangle as in the above example.   The other factor (which should be the height of the rectangle) then exposes the integrand and the range of integration.   Look at the following example.

EXAMPLE.   Find $\displaystyle \lim_{n \to \infty} \sum_{r=1}^{n} \frac{n^3}{(n^2 + r^2)^2}$.

Solution. $\displaystyle \sum_{r=1}^{n} \frac{n^3}{(n^2 + r^2)^2} = \frac{1}{n} \cdot \sum_{r=1}^{n} \frac{n^4}{(n^2 + r^2)^2}$

$$= \frac{1}{n} \cdot \sum_{r=1}^{n} \frac{1}{\left[1 + \left[\frac{r}{n}\right]^2\right]^2} \to \int_0^1 \frac{1}{(1 + x^2)^2} dx \quad \text{as} \quad n \to \infty.$$

by §239, because the integrand is continuous.   Putting $x = \tan y$ in the integral gives its value as $\int_0^{\pi/4} \cos^2 y \, dy = (2 + \pi)/8$.

EXAMPLE.   Show $\displaystyle \lim_{n \to \infty} \left\{ \frac{n}{(n+1)^2} + \frac{n}{(n+2)^2} + \ldots + \frac{n}{(2n)^2} \right\} = \frac{1}{2}$.

EXAMPLE. Show $\lim\limits_{n\to\infty} \frac{1}{n}(\sin\frac{\pi}{n} + \sin\frac{2\pi}{n} + \ldots + \sin\pi) = \frac{2}{\pi}$.

For further examples of this type see Exs.11.4 - 11.7 and 11.12.

247. THE USE OF INTEGRATION IN THE DEVELOPMENT OF A MACLAURIN EXPANSION

The following example illustrates this.

EXAMPLE. Prove that, for $x \epsilon [-1,1]$,

$$\tan^{-1}x = x - \frac{x^3}{3} + \frac{x^5}{5} - \ldots = \sum_{n=0}^{\infty} (-1)^n \frac{x^{2n+1}}{2n+1} .$$

Solution. Notice first that, for all $t\epsilon R$,

$$1 - t^2 + t^4 - t^6 + \ldots + (-1)^{n-1}t^{2n-2} = \frac{1-(-t^2)^n}{1+t^2} .$$

So $\frac{1}{1+t^2} = 1 - t^2 + t^4 - \ldots + (-1)^{n-1}t^{2n-2} + \frac{(-1)^n t^{2n}}{1+t^2}$.

So $\int_0^x \frac{dt}{1+t^2} = \int_0^x 1 - t^2 + t^4 - \ldots + (-1)^{n-1}t^{2n-2} dt + R_n(x)$,

$$\text{where } R_n(x) = \int_0^x \frac{(-1)^n t^{2n}}{1+t^2} dt.$$

So $[\tan^{-1}t]_0^x = [t - \frac{t^3}{3} + \frac{t^5}{5} - \ldots + (-1)^{n-1}\frac{t^{2n-1}}{2n-1}]_0^x + R_n(x)$,

i.e. $\tan^{-1}x = x - \frac{x^3}{3} + \frac{x^5}{5} - \ldots + (-1)^{n-1}\frac{x^{2n-1}}{2n-1} + R_n(x)$.

Look at $R_n(x)$ as $n \to \infty$. So,

$$|R_n(x)| = \left|\int_0^x \frac{t^{2n}}{1+t^2} dt\right| \le \left|\int_0^x t^{2n} dt\right| = \frac{|x|^{2n+1}}{2n+1} \quad ..(*)$$

$$\le \frac{1}{2n+1} \quad \text{for} \quad |x| \le 1$$

$$\to 0 \quad \text{as} \quad n \to \infty.$$

So $\tan^{-1}x = \sum_{n=0}^{\infty} (-1)^n \frac{x^{2n+1}}{2n+1}$ for $x \epsilon [-1,1]$, as required.

Notice here that if $|x| > 1$ then (*) in the above example tends to infinity as $n \to \infty$ so that we cannot conclude that the remainder tends to zero. Actually the series has radius of convergence equal to 1 so that there is no hope of it converging (let alone converging to $\tan^{-1}x$) for $|x| > 1$.

Compare this method with the method of developing the series for $\tan^{-1}x$ in §220.

EXAMPLE. It is given that, provided $t \neq -1$,

$$\frac{1}{1+t} = 1 - t + t^2 - \dots + (-1)^{n-1}t^{n-1} + \frac{(-1)^n t^n}{1+t} .$$

Prove that, for $x \in \,]-1,1]$,

$$\log(1+x) = \sum_{n=1}^{\infty} (-1)^{n-1} \frac{x^n}{n} .$$

IMPROPER INTEGRALS

248. Riemann integration was considered in §231 to §244 for the case of a **bounded** function over a **finite** interval. We now look at the possibility of extending these ideas to **unbounded** functions and/or **infinite** intervals.

249. INTEGRATION OVER AN INFINITE INTERVAL: DEFINITION OF THE INTEGRAL

Suppose that $f \in R[a,b]$ for all $b > a$. Then if $\lim_{b \to \infty} \int_a^b f(x)\,dx$ exists and has the value K (a **finite** value), we say that $\int_a^{\infty} f(x)\,dx$ **exists** (or **converges**) and has the value K. Otherwise $\int_a^{\infty} f(x)\,dx$ **does not exist** (or **diverges**).

So $\int_1^{\infty} 1/x^2\,dx$ exists and has the value 1. For

$$\int_1^{\infty} 1/x^2\,dx = \lim_{b \to \infty} \int_1^b 1/x^2\,dx = \lim_{b \to \infty} [-1/x]_1^b$$

$$= \lim_{b \to \infty} (1 - \frac{1}{b}) = 1 .$$

On the other hand, $\int_1^{\infty} 1/x\,dx$ does not exist. For

$$\int_1^{\infty} 1/x\,dx = \lim_{b \to \infty} \int_1^b 1/x\,dx = \lim_{b \to \infty} [\log x]_1^b$$

$$= \lim_{b \to \infty} \log b , \text{ which does not exist.}$$

Notice that, in general, $\int_1^\infty 1/x^k \, dx$ exists if and only if $k > 1$.

250. A COMPARISON TEST FOR THE CONVERGENCE OF INTEGRALS

RESULT. Suppose that $0 \le f(x) \le g(x)$ for all $x \in [N, \infty[$ and that $\int_a^\infty g(x) \, dx$ exists. Then, provided that $\int_a^b f(x) \, dx$ exists for each $b > a$, it follows that $\int_a^\infty f(x) \, dx$ exists. (Here N is any number exceeding a.)

Proof. Routine.

The following example illustrates the use of this result.

EXAMPLE. Show that for all values of $m, n \in R$ the integral $\int_1^\infty e^{-x} x^m (\log x)^n \, dx$ converges.

Solution. By the hierarchy result of §209, $e^{-x} x^{m+2} (\log x)^n \to 0$ as $x \to \infty$. So there exists N such that $0 < e^{-x} x^{m+2} (\log x)^n < 1$ for all $x \ge N$. So $0 < e^{-x} x^m (\log x)^n < 1/x^2$ for all $x \ge N$. Since $\int_1^\infty 1/x^2 \, dx$ exists (§249), the above result guarantees the existence of the given integral.

251. The integral test of §182 can be stated as:

RESULT. Let f be a positive decreasing function on $[a, \infty[$. Then $\int_a^\infty f(x) \, dx$ converges or diverges according as $\sum f(n)$ converges or diverges.

252. INTEGRATION OF AN UNBOUNDED FUNCTION: DEFINITION OF THE INTEGRAL

Suppose that $f(x) \to \infty$ as $x \to b-$. Then the definition of §233 cannot be used to consider the Riemann integrability of f on $[a,b]$ because f is unbounded. However, suppose that $f \in R[a,c]$ for every c with $a < c < b$. Then if we find that

$\lim_{c \to b-} \int_a^c f(x)\, dx$ exists and has the value K (a _finite_ value), then we say that $\int_a^b f(x)\, dx$ _exists_ (or _converges_) and has the value K. Otherwise we say that $\int_a^b f(x)\, dx$ _does not exist_ (or _diverges_). Similarly for the integration of a function which tends to $\pm\infty$ at any finite number of points in $[a,b]$.

So $\int_0^1 1/\sqrt{x}\, dx$ exists and has the value 2. For

$$\int_0^1 1/\sqrt{x}\, dx = \lim_{b \to 0+} \int_b^1 1/\sqrt{x}\, dx = \lim_{b \to 0+} (2 - 2\sqrt{b}) = 2.$$

On the other hand, $\int_0^1 1/x\, dx$ does not exist. For

$$\int_0^1 1/x\, dx = \lim_{b \to 0+} \int_b^1 1/x\, dx = \lim_{b \to 0+} (- \log b), \text{ which}$$

does not exist.

N.B.   (i) $\int_0^1 1/x^k\, dx$ exists if and only if $k < 1$.

(ii)   The two examples worked out above illustrate that for unbounded functions the fact that $\int_0^1 f(x)\, dx$ exists does not guarantee the existence of $\int_0^1 (f(x))^2\, dx$: take $f(x) = 1/\sqrt{x}$ on $[0,1]$.

253.   ANOTHER TEST FOR THE CONVERGENCE OF INTEGRALS

RESULT.   Suppose that $f(x) \to \infty$ as $x \to b-$. Suppose that g is Riemann integrable on $[a,b]$ and that $g(x)$ tends to a _non-zero_ limit K as $x \to b-$. Then $\int_a^b f(x) g(x)\, dx$ converges if and only if $\int_a^b f(x)\, dx$ converges.

Proof.   Without loss of generality suppose that $K > 0$. If we take $\varepsilon = \frac{1}{2}K$ in the definition of a limit, we find a number $\delta > 0$ such that, for all $x \in [b-\delta, b[$, we have both that $\frac{1}{2}K < g(x) < \frac{3}{2}K$ and that $f(x) > 0$. So $\frac{1}{2}K f(x) \le f(x) g(x) \le \frac{3}{2}K f(x)$ for all $x \in [b-\delta, b[$.

So $\frac{1}{2}K \int_{b-\delta}^b f(x)\, dx \le \int_{b-\delta}^b f(x) g(x)\, dx \le \frac{3}{2}K \int_{b-\delta}^b f(x)\, dx$.

The result follows from this.

Notice how this last result relates the existence of $\int_a^b f(x)g(x)\,dx$ to the existence of $\int_a^b K\,f(x)\,dx$. In this situation we say that $f(x)g(x)$ <u>behaves like</u> $K\,f(x)$ as $x \to b-$.

<u>EXAMPLE</u>.   Show that $\int_0^1 x^{p-1}(1-x)^{q-1}\,dx$ exists if and only if p and q are both positive.

<u>Solution</u>.   As $x \to 0+$, the integrand behaves like x^{p-1} and $\int_0^1 x^{p-1}\,dx$ converges if and only if $p > 0$ by (i) of §252.

Similarly, as $x \to 1-$, the integrand behaves like $(1-x)^{q-1}$ and $\int_0^1 (1-x)^{q-1}\,dx$ converges if and only if $q > 0$.

Intersect the two conditions to see that the given integral converges if and only if $p > 0$ and $q > 0$.

<u>EXAMPLE</u>.   Show that $\int_0^1 x \log x\,dx$ exists as a Riemann integral (i.e. under the definition of §233).

Show that $\int_0^1 e^{-x} x^{-\frac{1}{2}}\,dx$, $\int_0^1 \log x\,dx$ and $\int_0^1 e^{-x}\log x\,dx$ all exist as improper Riemann integrals (i.e. under the definition of §252).

Show that $\int_0^1 e^{-x}/x\,dx$ does <u>not</u> exist.

BETA AND GAMMA FUNCTIONS

254.   DEFINITIONS OF THESE FUNCTIONS AS INTEGRALS

The values taken by these functions are given as the values of the following two integrals:

(i) $B(p,q) = \int_0^1 x^{p-1}(1-x)^{q-1}\,dx$,   ($p > 0$ and $q > 0$),

(ii) $\Gamma(k) = \int_0^\infty x^{k-1}e^{-x}\,dx$,   ($k > 0$).

Demanding that these integrals exist forces the given restrictions on p, q and k: for (i) see §253;

for (ii) see §252 as x → 0+ and §250 with n = 0 as
x → ∞.

The Beta function has two other useful forms:

$$(iii) \quad B(p,q) = 2 \int_0^{\frac{1}{2}\pi} \sin^{2p-1}y \cos^{2q-1}y \, dy ,$$

$$(iv) \quad B(p,q) = \int_0^\infty \frac{y^{p-1}}{(1+y)^{p+q}} \, dy .$$

To get (iii) put $x = \sin^2 y$ in (i). For (iv) put
$x = y/(1+y)$ in (i).

Many integrals in Probability and Statistics in
connection with the probability density functions of
normal, t, χ^2 and Gamma random variables can be
reduced to integrals of the above four types.

255. PROPERTIES OF THE BETA AND GAMMA FUNCTIONS

A. $\Gamma(1) = 1$, $\Gamma(2) = 1$, $\Gamma(3) = 2$ and in general
$\Gamma(n) = (n-1)!$ for every n∈N.

B. $\Gamma(k) = (k-1)\,\Gamma(k-1)$ for all k > 1.   Repeated
application of this reduces the calculation of
$\Gamma(k)$ to that of $\Gamma(k-p)$ where p∈N and
$0 < k-p \leq 1$. E.g. $\Gamma(7/2) = \frac{5}{2} \cdot \frac{3}{2} \cdot \frac{1}{2} \Gamma(\frac{1}{2})$.

C.   For $0 < k < 1$, $\Gamma(k) \cdot \Gamma(1-k) = \pi/\sin k\pi$.
E.g. $\Gamma(1/4) \cdot \Gamma(3/4) = \sqrt{2}\pi$.

D. $\Gamma(\frac{1}{2}) = \sqrt{\pi}$.   (A special case of C)

E. $B(p,q) = \dfrac{\Gamma(p)\Gamma(q)}{\Gamma(p+q)}$.

Of these, A and B are easy to prove with integration
by parts.   C,D and E are harder and we leave them.
Notice that E allows every Beta function to be expressed
in terms of Gamma functions and that B expresses every
Gamma function in terms of $\Gamma(p)$ for some p with
$0 < p \leq 1$.

256.   In attempting to reduce a given integral to a Beta

or Gamma function, decide at the outset which of
the four forms in §254 you are aiming for.   Then make a
change of variable to achieve it.   Looking at the
limits of integration can help both in deciding what to
aim for and in deciding the change of variable.

EXAMPLE.   By making suitable changes of variables express each of the following integrals as a Beta function or a Gamma function and hence show that

(a) $\displaystyle\int_0^\infty e^{-x^2}\,dx = \tfrac{1}{2}\sqrt{\pi}$,   (b) $\displaystyle\int_0^1 \sqrt{(x(1-x))}\,dx = \dfrac{\pi}{8}$,

(c) $\displaystyle\int_0^2 x^{5/2}(4-x^2)^{1/4}\,dx = \dfrac{3\pi\sqrt{2}}{4}$,   (d) $\displaystyle\int_0^\infty \dfrac{dx}{1+x^4} = \dfrac{\pi\sqrt{2}}{4}$,

(e) $\displaystyle\int_0^{\frac{1}{2}\pi} \sin^3 y \cos^7 y \, dy = \dfrac{1}{40}$,   (f) $\displaystyle\int_0^\infty \dfrac{t}{2+t^4}\,dt = \dfrac{\pi\sqrt{2}}{8}$,

(g) $\displaystyle\int_0^\infty \dfrac{t^2}{(1+t^2)^3}\,dt = \dfrac{\pi}{16}$,   (h) $\displaystyle\int_0^{\frac{1}{2}\pi} \sqrt{(\tan y)}\,dy = \dfrac{\pi\sqrt{2}}{2}$.

[Hints: (a) $u = x^2$, (b) already in form (i) of §254, (c) $x^2 = 4u$, (d) $x^4 = u$, (e) already in form (iii), (f) $t^4 = 2u$, (g) $t^2 = u$, (h) $\tan y = \sin y/\cos y$.]

257.   The following integral is of frequent occurrence in certain areas.

RESULT. $\displaystyle\int_0^\infty x^{\alpha-1} e^{-kx^\beta}\,dx = \dfrac{\Gamma(\alpha/\beta)}{\beta\, k^{\alpha/\beta}}$, where k, α, β are

all positive.

Proof.   Put $u = kx^\beta$ in the integral.

258.   From the properties of the Beta and Gamma functions the following rule can be derived concerning the integral (iii) of §254.

RESULT.   For non-negative integers m and n,

$$\int_0^{\frac{1}{2}\pi} \sin^m x \cos^n x \, dx = \frac{(m-1)(m-3)\ldots(n-1)(n-3)\ldots}{(m+n)(m+n-2)(m+n-4)\ldots}\cdot K$$

where $K = 1$ unless m and n are both even in which case $K = \tfrac{1}{2}\pi$. The dots in the product indicate that the factors continue till 1 or 2 is reached. If $m = 0$ or $m = 1$, none of the factors $(m-1)$, $(m-3)$ appears etc.

EXAMPLE. $\displaystyle\int_0^{\frac{1}{2}\pi} \sin^6 x \cos^5 x \, dx = \dfrac{5.3.1.4.2}{11.9.7.5.3.1} = \dfrac{8}{693}$.

$\displaystyle\int_0^{\frac{1}{2}\pi} \sin^8 x \, dx = \dfrac{7.5.3.1}{8.6.4.2}\cdot\tfrac{1}{2}\pi = \dfrac{35\pi}{256}$.

259. DIFFERENTIATION UNDER THE INTEGRAL SIGN

<u>RESULT</u>. Suppose that the equation $\quad g(a) = \displaystyle\int_{u(a)}^{v(a)} f(x)\,dx$

defines g as a function of a. (Here u and v are differentiable functions and f is Riemann integrable between all possible values $u(a)$ and $v(a)$.) Then

$$g'(a) = f(v(a))v'(a) - f(u(a))u'(a).$$

<u>Proof</u>. Let F be an indefinite integral of f. Then $g(a) = F(v(a)) - F(u(a))$. So

$$g'(a) = F'(v(a))v'(a) - F'(u(a))u'(a)$$
$$= f(v(a))v'(a) - f(u(a))u'(a), \text{ as stated.}$$

As an example of this result notice that

$$\frac{d}{da}\left(\int_{0}^{\sqrt{a}} e^{-\frac12 x^2}\,dx \right) = \frac{1}{2\sqrt{a}}\, e^{-\frac12 a}.$$

In probability theory this shows that if the random variable X has a normal distribution with $E(X) = 0$ and $V(X) = 1$ then X^2 has a chi-squared distribution with one degree of freedom.

As a generalisation of the above result notice that under suitable conditions

$$\text{if}\quad h(a) = \int_{u(a)}^{v(a)} f(x,a)\,dx, \quad\text{then}$$

$$h'(a) = \int_{u(a)}^{v(a)} \frac{\partial f}{\partial a}\,dx + f(v(a),a)v'(a) - f(u(a),a)u'(a).$$

260. RIEMANN STIELTJES INTEGRATION

Let f be a bounded function and let g be an increasing function on $[a,b]$. As a generalisation of the situation of §232 and §236 take

$$U(f,g:D) = \sum_{k=1}^{n} M_k (g(c_k) - g(c_{k-1})),$$

$$L(f,g:D) = \sum_{k=1}^{n} m_k (g(c_k) - g(c_{k-1})),$$

$$S(f,g:D,e) = \sum_{k=1}^{n} f(e_k)(g(c_k) - g(c_{k-1}),$$

where D is a dissection of $[a,b]$ and M_k, m_k and e_k

are as for Riemann integration in §232 and §236. (The Riemann integral is the particular case with g(x) = x.)

In the case when sup L(f,g:D) and inf U(f,g:D) are
D   D
equal we say that f is <u>Riemann Stieltjes integrable</u> <u>with respect to g on [a,b]</u> and we write $\int_a^b f\,dg$ for their common value.

Equivalently, as in §236, if lim S(f,g:D,e) = A for
D
all choices of the points e_k then f is Riemann Stieltjes integrable with respect to g and $\int_a^b f\,dg = A$.

261.   For Riemann Stieltjes integrals it is possible to prove (a) heredity results (on integrability of sums, products etc of integrable functions), and (b) results giving sufficient conditions on f and g to ensure that $\int_a^b f\,dg$ exists.   We do not pursue this.

262.   REDUCTION OF RIEMANN STIELTJES INTEGRALS TO RIEMANN INTEGRALS

The following result shows that this is possible in certain cases.

<u>RESULT</u>.   If $\int_a^b f\,dg$ exists and g is differentiable on [a,b] (as well as being monotonic), then

$$\int_a^b f\,dg = \int_a^b f(x)\,g'(x)\,dx.$$

We do not prove this result but as some indication notice that if g is differentiable we can apply the mean-value theorem to see that

$$S(f,g:D,e) = \sum_{k=1}^n f(e_k)(g(c_k) - g(c_{k-1}))$$

$$= \sum_{k=1}^n f(e_k)g'(\xi_k)(c_k - c_{k-1}), \text{ where } \xi_k \in]c_{k-1}, c_k[.$$

263.   AN APPLICATION OF THE RIEMANN STIELTJES INTEGRAL IN PROBABILITY

The <u>expected value</u> of a random variable X is defined as <u>follows</u>:

(a)   If X is a <u>discrete</u> random variable which takes the values $x_1, x_2, x_3, \ldots$ with the probabilities

P_1, P_2, P_3, ... then $E(X) = \sum x_i p_i$.

(b) If X is a <u>continuous</u> random variable with probability density function g, then $E(X) = \int_{-\infty}^{\infty} x\, g(x)\, dx$.

Using the Riemann Stieltjes integral we can give a single definition of $E(X)$ which covers both cases as follows.

Let X be a random variable and let G be its cumulative distribution function. Then G is an increasing function on R. In particular in the situation of case (b) above

$$G(b) = \int_{-\infty}^{b} g(x)\, dx .$$

In both case (a) and case (b) we can define $E(X)$ by

$$E(X) = \int_{-\infty}^{\infty} f\, dG, \quad \text{where}\quad f(x) = x.$$

To see that this definition coincides with that in (b) for the case of a continuous probability density function g, notice that $G' = g$ and apply the result of §262 to $\int_{-\infty}^{\infty} f\, dG$.

This definition also applies to random variables that are partly discrete and partly continuous as in the following example.

<u>EXAMPLE</u>. The random variable X takes the value 3 with probability $\frac{3}{4}$ and otherwise it is uniformly distributed on the interval [5,6]. Show that $E(X) = \frac{29}{8}$.

EXAMPLES 11

1. Prove that the function f defined on [0,1] by

$f(x) = 2$ (x irrational), $f(x) = 1$ (x rational)

is <u>not</u> Riemann integrable on [0,1]. Give an example to show that even if $g \notin R[0,1]$, it is still possible that $g^2 \in R[0,1]$ and $|g| \in R[0,1]$.

2. If $f \in R[0,1]$ and $f(x) \geq k > 0$ for all $x \in [0,1]$, prove that $1/f \in R[0,1]$.

3. From area considerations and §258 show that
$$\int_{0}^{\pi} \sin^5 x \, \cos^2 x \, dx = \frac{16}{105} , \qquad \int_{0}^{\pi} \sin^5 x \, \cos^3 x \, dx = 0 .$$

4.   Use Riemann integration to prove that as $n \to \infty$

(i) $\left[\dfrac{1}{2n+1} + \dfrac{1}{2n+2} + \ldots + \dfrac{1}{3n}\right] \to \log \dfrac{3}{2}$,

(ii) $n\left[\dfrac{1}{(2n+1)^2} + \dfrac{1}{(2n+2)^2} + \ldots + \dfrac{1}{(3n)^2}\right] \to \dfrac{1}{6}$,

(iii) $n\left[\dfrac{1}{n^2+1^2} + \dfrac{1}{n^2+2^2} + \ldots + \dfrac{1}{2n^2}\right] \to \dfrac{\pi}{4}$.

5.   Notice in the results of Ex.1.14 that the coefficients of the term of highest degree on the right-hand side are 1/2, 1/3, 1/4. Generalise this by proving that, for $p > -1$,

$$\lim_{n \to \infty} \frac{1^p + 2^p + \ldots + n^p}{n^{p+1}} = \frac{1}{p+1} .$$

6.   Use Riemann integration to show that

$$\lim_{n \to \infty} \frac{1}{n}[\log(1+\tfrac{1}{n}) + \log(1+\tfrac{2}{n}) + \ldots + \log 2] = 2\log 2 - 1.$$

Deduce that $\dfrac{1}{n}\left[(n+1)(n+2)\ldots(2n)\right]^{1/n} \to \dfrac{4}{e}$ as $n \to \infty$.

7.   Prove that

$$\lim_{n \to \infty} \frac{1}{n^2}\left[\log\frac{1}{n} + 2\log\frac{2}{n} + \ldots + n\log\frac{n}{n}\right] = -\frac{1}{4} .$$

Deduce that $\dfrac{1}{\sqrt{n}}\left[1^1 2^2 3^3 \ldots n^n\right]^{1/n^2} \to e^{-1/4}$ as $n \to \infty$.

8.   (Beta and Gamma functions)   Show that

(a) $\displaystyle\int_0^3 \sqrt{(3x - x^2)}\, dx = \dfrac{9\pi}{8}$,   (b) $\displaystyle\int_{-\infty}^{\infty} e^{-\frac{1}{2}x^2}\, dx = \sqrt{(2\pi)}$,

(c) $\displaystyle\int_0^1 \dfrac{x\, dx}{\sqrt{(1-x^4)}} = \dfrac{\pi}{4}$,   (d) $\displaystyle\int_0^{\frac{1}{2}\pi} \sqrt{(\sin x)}\, dx = \left(\dfrac{2}{\pi}\right)^{\frac{1}{2}} (\Gamma(\tfrac{3}{4}))^2$,

(e) $\displaystyle\int_1^{\infty} (\log x)^3 x^{-2}\, dx = 6$,   (f) $\displaystyle\int_0^{\infty} \dfrac{dx}{1 + 8x^3} = \dfrac{\pi\sqrt{3}}{9}$.

9.   Let a and b be positive numbers and let

$$I = \int_0^{\frac{1}{2}\pi} \frac{\cos^2 x\, dx}{a^2\cos^2 x + b^2\sin^2 x} , \quad J = \int_0^{\frac{1}{2}\pi} \frac{\sin^2 x\, dx}{a^2\cos^2 x + b^2\sin^2 x} .$$

Show that $a^2 I + b^2 J = \frac{1}{2}\pi$, $I + J = \dfrac{\pi}{2ab}$. Hence find I, J.

10. (Cauchy's inequality for integrals)   Let
f, g ε C[a,b].   Notice that it can be proved along
similar lines to the proof in §277 that

$$\left| \int_a^b f g \, dx \right| \leq \left(\int_a^b f^2 \, dx \right)^{\frac{1}{2}} \left(\int_a^b g^2 \, dx \right)^{\frac{1}{2}}.$$

11.   Taking the definition of   log x   suggested in
§208, prove that   log(xy) = log x + log y   for all
positive   x   and   y.

12.   Use integration to show that

$$\lim_{n\to\infty} \frac{1}{n^2}\left[(n+1)\log(1+\tfrac{1}{n}) + (n+2)\log(1+\tfrac{2}{n}) + \ldots + 2n \log 2 \right]$$

$$= 2 \log 2 - \frac{3}{4}.$$

Deduce that $\lim_{n\to\infty} n^{-3/2}[(n+1)^{n+1}(n+2)^{n+2}\ldots(2n)^{2n}]^{1/n^2}$

has the value $4e^{-3/4}$.   [You will need the fact that
$1 + 2 + 3 + \ldots + n = \tfrac{1}{2}n(n+1)$.]

13.   Use the fact that $x^{1/4}\log x \to 0$ as $x \to 0+$ to
show that there exists   k > 0   such that $|\log x| \leq x^{-1/4}$
for $0 < x \leq k$.   Use this fact to show that
$\int_0^1 x^{-\frac{1}{2}}|\log x| \, dx$ exists as an improper Riemann integral.

14.   Let   a, b   be non-zero numbers.   Use integration
by parts to show that

$$b\int e^{ax}\cos bx \, dx = e^{ax}\sin bx - a\int e^{ax}\sin bx \, dx,$$
$$b\int e^{ax}\sin bx \, dx = -e^{ax}\cos bx + a\int e^{ax}\cos bx \, dx.$$

Deduce that

$$\int e^{ax}\cos bx \, dx = \frac{e^{ax}}{a^2+b^2}\left[a \cos bx + b \sin bx\right],$$

$$\int e^{ax}\sin bx \, dx = \frac{e^{ax}}{a^2+b^2}\left[a \sin bx - b \cos bx\right].$$

Obtain these last two results formally by writing
$e^{ax}(\cos bx + i \sin bx) = e^{(a+ib)x}$ and integrating this.

XII.   MORE ON CONVERGENCE OF SEQUENCES AND SERIES

264.   The <u>ratio</u> a_{n+1}/a_n of consecutive terms of a sequence or series can give information about convergence.   As examples of this take the method of §65 for sequences and the ratio test of §151 for series. This ratio appears again in several of the following results.

265.   O NOTATION FOR SEQUENCES

The O notation was used in connection with truncation of series in §193.   It can be used in connection with the behaviour of sequences as $n \to \infty$.

<u>Definition</u>.   Let $\{f(n)\}$ be a real sequence and let $\{g(n)\}$ be a positive sequence.   Then $f(n) = O(g(n))$ as $n \to \infty$ means:

$\exists \, K > 0$ and $\exists \, X \varepsilon R$ such that $|f(n)| \leq K \, g(n)$ $\forall n > X.$

<u>EXAMPLES</u>.   (i) $\dfrac{2}{n} - \dfrac{6}{n^2} + \dfrac{3}{n^5} = O\!\left(\dfrac{1}{n}\right)$ as $n \to \infty$.   For

$\left|\dfrac{2}{n} - \dfrac{6}{n^2} + \dfrac{3}{n^5}\right| \leq \dfrac{2}{n} + \dfrac{6}{n} + \dfrac{3}{n} = \dfrac{11}{n}$ for all $n > 1.$

(ii) $\log\!\left(1 + \dfrac{1}{n}\right) = \dfrac{1}{n} - \dfrac{1}{2n^2} + O\!\left(\dfrac{1}{n^3}\right)$ for all $n \geq 2,$

on using the power series for $\log(1 + x)$ from §212.

(iii) $\left(1 - \dfrac{1}{n}\right)^{-1} = 1 + \dfrac{1}{n} + O\!\left(\dfrac{1}{n^2}\right)$ for all $n \geq 2$ on

using the second series of §176.

266.   Notice the technicalities of manipulating the O notation in the following example.   The binomial theorem is also useful in this situation.

<u>EXAMPLE</u>.   Express $\left(\dfrac{n+2}{n+5}\right)\left(\dfrac{n}{n+1}\right)^{\frac{1}{2}}$ in the form $1 + \dfrac{a}{n} + O\!\left(\dfrac{1}{n^2}\right).$

<u>Solution</u>. $\left(\dfrac{n+2}{n+5}\right)\left(\dfrac{n}{n+1}\right)^{\frac{1}{2}} = \left(1 + \dfrac{2}{n}\right)\left(1 - \dfrac{5}{n} + O\!\left(\dfrac{1}{n^2}\right)\right)\left(1 + \dfrac{1}{n}\right)^{-\frac{1}{2}}$

$= \left(1 + \dfrac{2}{n}\right)\left(1 - \dfrac{5}{n} + O\!\left(\dfrac{1}{n^2}\right)\right)\left(1 - \dfrac{1}{2n} + O\!\left(\dfrac{1}{n^2}\right)\right) = 1 - \dfrac{7}{2n} + O\!\left(\dfrac{1}{n^2}\right).$

267. ANOTHER USEFUL NOTATION

Definition. Let $\{f(n)\}$ and $\{g(n)\}$ be real sequences. Then $f(n) \sim g(n)$ as $n \to \infty$ means that $f(n)/g(n) \to 1$ as $n \to \infty$.

(This definition extends easily to real functions, e.g. $f(x) \sim g(x)$ as $x \to \infty$.)

EXAMPLES. (i) $3n^2 + 7n + 1 \sim 3n^2$ as $n \to \infty$.

(ii) $4n + 5 \log n \sim 4n$ as $n \to \infty$.

(iii) $\cosh x \sim \frac{1}{2}e^x$ as $x \to \infty$.

CONVERGENCE OF SEQUENCES

268. SOME SEQUENCES WITH NON-ZERO LIMITS

The following result is used in the proofs of the results in §269 - §272. In all these results the ratio of consecutive terms of a sequence appears.

RESULT. Let $\{f(n)\}$ be a positive sequence such that

$$\frac{f(n+1)}{f(n)} = 1 + O\left(\frac{1}{n^2}\right) \quad \text{as} \quad n \to \infty.$$

Then $\{f(n)\}$ tends to a non-zero limit as $n \to \infty$.

Proof. Notice first that we can write

$$\frac{f(n+1)}{f(n)} = 1 + x_n ,$$

where $|x_n| < 1$ for all sufficiently large n. For such n, we can therefore expand $\log(1 + x_n)$ using the series for $\log(1 + x)$ from §212. Truncating this series at the first term gives constants K, Y such that

$$\left|\log\left(\frac{f(n+1)}{f(n)}\right)\right| = \left|\log\left(1 + O\left(\frac{1}{n^2}\right)\right)\right| \le \frac{K}{n^2} \quad \text{for all} \quad n > Y,$$

i.e. $|\log(f(n+1)) - \log(f(n))| \le \frac{K}{n^2}$ for all $n > Y$.

So $\sum |\log(f(n+1)) - \log(f(n))|$ converges since $\sum \frac{1}{n^2}$ converges. So $\sum (\log(f(n+1)) - \log(f(n)))$ converges by the result of §156. Taking the nth partial sum of this series deduce that $(\log(f(n+1)) - \log(f(1)))$ tends to a limit S as $n \to \infty$. So $\log(f(n+1))$ tends to $S + \log(f(1)) = L$, say as $n \to \infty$. So it follows that $f(n) \to e^L$ as $n \to \infty$. Also $e^L > 0$ because the real exponential takes only positive values. So the result is proved.

269. THE PRODUCT DEFINITION OF THE GAMMA FUNCTION

For every $x > 0$, define a sequence $\{g_x(n)\}$ by

$$g_x(n) = \frac{n! \, n^{x-1}}{x(x+1)(x+2)\ldots(x+n-1)} \quad \text{for all} \quad n \varepsilon N.$$

Then $\dfrac{g_x(n+1)}{g_x(n)} = \dfrac{(n+1)(n+1)^{x-1}}{n^{x-1}(x+n)} = \dfrac{(n+1)^x}{n^x\left[1 + \dfrac{x}{n}\right]}$

$$= \left(1 + \frac{1}{n}\right)^x \left(1 + \frac{x}{n}\right)^{-1} = \left(1 + \frac{x}{n} + O\left(\frac{1}{n^2}\right)\right)\left(1 - \frac{x}{n} + O\left(\frac{1}{n^2}\right)\right) = 1 + O\left(\frac{1}{n^2}\right).$$

By §268 we can then conclude that $\{g_x(n)\}$ tends to a non-zero limit as $n \to \infty$. It turns out that

$$\lim_{n \to \infty} g_x(n) = \Gamma(x),$$

where Γ denotes the Gamma function of §254.

EXAMPLE. Show that $\sqrt{n} \cdot \dfrac{1.3.5\ldots(2n-1)}{2.4.6\ldots(2n)}$ tends to a

positive limit A as $n \to \infty$. Deduce that

$$\frac{1.3.5\ldots(2n-1)}{2.4.6\ldots(2n)} \sim \frac{A}{\sqrt{n}} \quad \text{as} \quad n \to \infty.$$

(Actually from above it follows that $A = 1/\Gamma(\tfrac{1}{2}) = 1/\sqrt{\pi}$.)

270. STIRLING'S APPROXIMATION FOR N!

This has already been touched on in Ex.8.3. The result is that

$$n! \sim \sqrt{(2\pi n)} \left(\frac{n}{e}\right)^n \quad \text{as} \quad n \to \infty.$$

The method of §268 can be used to reach this, apart from the value of the constant $\sqrt{(2\pi)}$: the following example suggests how to proceed.

EXAMPLE. For each $n \varepsilon N$, define $f(n) = n! \, e^n \, n^{-(n+\frac{1}{2})}$.
Show that $\dfrac{f(n+1)}{f(n)} = 1 + O\left(\dfrac{1}{n^2}\right)$ and deduce that $\{f(n)\}$
tends to a positive limit A as $n \to \infty$. Deduce that
$n! \sim A\sqrt{n}\left(\dfrac{n}{e}\right)^n$ as $n \to \infty$.

271. THE INFINITE PRODUCT FOR SIN X AND WALLIS'S PRODUCT FOR $\frac{1}{2}\pi$

EXAMPLE. For every $x \varepsilon R$, define a sequence $\{f_x(n)\}$ by

$$f_x(n) = x\left(1 - \frac{x^2}{\pi^2}\right)\left(1 - \frac{x^2}{2^2\pi^2}\right) \cdots \left(1 - \frac{x^2}{n^2\pi^2}\right)$$

for each $n \varepsilon N$. Using §268 show that, for each $x \neq k\pi$
$(k \varepsilon Z)$, $\{f_x(n)\}$ tends to a non-zero limit $f(x)$ as
$n \to \infty$. Prove that, for $x = k\pi$ $(k \varepsilon Z)$, $f_x(n) \to 0$ as
$n \to \infty$. So we can define $f(k\pi) = 0$ for each $k \varepsilon Z$.

It actually turns out that $f(x) = \sin x$ for all
$x \varepsilon R$, though we cannot prove it here. Assuming this and
by putting $x = \frac{1}{2}\pi$, deduce that

$$\frac{1}{2}\pi = \lim_{n\to\infty} \frac{2^2 . 4^2 . 6^2 \ldots (2n)^2}{1.3.3.5.5.7 \ldots (2n-1)(2n+1)} \, .$$

(This last result is Wallis's product.)

Solution. Let $x \neq k\pi$ $(k \varepsilon Z)$ and notice that

$$\frac{f_x(n+1)}{f_x(n)} = 1 - \frac{x^2}{(n+1)^2 \, \pi^2} = 1 - \frac{x^2}{n^2 \pi^2 \left(1 + \frac{1}{n}\right)^2} = 1 + O\left(\frac{1}{n^2}\right).$$

A little thought shows that according to the value of x
the terms of the sequence $\{f_x(n)\}$ are either eventually
positive or eventually negative. In the positive case
we apply §268 directly, in the negative case we apply
§268 to $\{-f_x(n)\}$, but in each case we conclude that
$\{f_x(n)\}$ tends to a non-zero limit as $n \to \infty$.

Now let $x = k\pi$ $(k \varepsilon Z)$ and notice that $f_x(n) = 0$
for all $n \geq |k|$. So then $f_x(n) \to 0$ as $n \to \infty$.

So we have a limit function f with the properties
that $f_x(n) \to f(x)$ as $n \to \infty$, and $f(x) = 0$ if and only

if $x = k\pi$ $(k\varepsilon Z)$. [Notice at this point that this shows that f certainly has the same zeros as the sine function though we are not in a position to <u>prove</u> that $f(x) = \sin x$.]

So we <u>assume</u> that

$$\sin x = \lim_{n \to \infty} \; x\left(1 - \frac{x^2}{\pi^2}\right)\left(1 - \frac{x^2}{2^2\pi^2}\right)\cdots\left(1 - \frac{x^2}{n^2\pi^2}\right),$$

and put $x = \tfrac{1}{2}\pi$. This gives

$$1 = \lim_{n \to \infty} \; \tfrac{1}{2}\pi\left(1 - \frac{1}{2^2}\right)\left(1 - \frac{1}{4^2}\right)\cdots\left(1 - \frac{1}{(2n)^2}\right)$$

i.e. $$1 = \lim_{n \to \infty} \; \tfrac{1}{2}\pi\left(\frac{2^2 - 1}{2^2}\right)\left(\frac{4^2 - 1}{4^2}\right)\cdots\left(\frac{(2n)^2 - 1}{(2n)^2}\right).$$

So $$\lim_{n \to \infty} \; \frac{2^2 . 4^2 . 6^2 \ldots (2n)^2}{1.3.3.5.5.7\ldots(2n-1)(2n+1)} = \tfrac{1}{2}\pi.$$

Notice that the above result for $\sin x$ may be expressed in infinite product notation as follows:

$$\frac{\sin x}{x} = \prod_{n=1}^{\infty}\left(1 - \frac{x^2}{n^2\pi^2}\right)$$

with suitable conventions about the meaning when x is an integral multiple of π.

272. SOME SEQUENCES THAT CONVERGE TO ZERO

The following result makes a statement about the individual terms of a sequence from an examination of the ratio of consecutive terms. It is useful in connection with the tests of Dirichlet and Gauss in §274 and §276.

<u>RESULT</u>. Let $\{f(n)\}$ be a positive sequence such that

$$\frac{f(n+1)}{f(n)} = 1 - \frac{k}{n} + O\left(\frac{1}{n^2}\right) \qquad \text{as} \quad n \to \infty,$$

where $k > 0$. Then $f(n) \to 0$ as $n \to \infty$. Furthermore, there exists a positive constant A such that $f(n) \sim A/n^k$ as $n \to \infty$.

<u>Proof</u>. Let $g(n) = n^k f(n)$ for all $n\varepsilon N$. Then

$$\frac{g(n+1)}{g(n)} = \frac{(n+1)^k}{n^k} \cdot \frac{f(n+1)}{f(n)}$$

$$= \left(1 + \frac{k}{n} + O\left(\frac{1}{n^2}\right)\right)\left(1 - \frac{k}{n} + O\left(\frac{1}{n^2}\right)\right) = 1 + O\left(\frac{1}{n^2}\right).$$

From §268 deduce that there exists a positive number A such that $g(n) \to A$ as $n \to \infty$. So $n^k f(n) \to A$ as $n \to \infty$. So $f(n) \sim A/n^k$ as $n \to \infty$. As a consequence of this, since $k > 0$, $f(n) \to 0$ as $n \to \infty$.

EXAMPLE. Find constants α and β such that

$$\frac{1.4.7\ldots(3n-2)}{2.5.8\ldots(3n-1)} \sim An^{-\alpha} \quad \text{as} \quad n \to \infty$$

and

$$\frac{5.11.17\ldots(6n-1)}{7.13.19\ldots(6n+1)} \sim Bn^{-\beta} \quad \text{as} \quad n \to \infty$$

for some positive constants A and B. (Use this last result.)

CONVERGENCE OF A POWER SERIES
ON ITS CIRCLE OF CONVERGENCE

273. In §192 the examples $\sum z^n$, $\sum z^n/n$ and $\sum z^n/n^2$ are given to show that a power series can converge at none, some or all of the points on its circle of convergence. For the series $\sum z^n/n$, methods given earlier easily settle the behaviour of the series for $z = -1$ and $z = 1$, but they are not sufficiently delicate to decide whether this series converges for other points with $|z| = 1$. The tests of Dirichlet (§274) and Gauss (§276) settle the behaviour of $\sum z^n/n$ and of many other power series at all points on their circles of convergence.

274. DIRICHLET'S TEST

RESULT. (Dirichlet's test) Suppose that $\{b_n\}$ is a real sequence decreasing to zero. Suppose also that $\sum a_n$ is a series of real or complex terms with bounded partial sums. Then $\sum a_n b_n$ converges.

Proof. Let $s_n = \sum_{k=1}^{n} a_k$ and $t_n = \sum_{k=1}^{n} a_k b_k$ and by hypothesis let $|s_n| \leq C$ for all $n \in \mathbb{N}$.

Then, for $m > n$,

$$t_m - t_n = a_{n+1}b_{n+1} + a_{n+2}b_{n+2} + \ldots + a_m b_m$$

$$= (s_{n+1}-s_n)b_{n+1} + (s_{n+2}-s_{n+1})b_{n+2} + \ldots + (s_m - s_{m-1})b_m$$

$$= s_{n+1}(b_{n+1}-b_{n+2}) + s_{n+2}(b_{n+2}-b_{n+3}) + \ldots$$

$$\ldots + s_{m-1}(b_{m-1}-b_m) - s_n b_{n+1} + s_m b_m.$$

So

$$|t_m - t_n| \leq C(b_{n+1}-b_{n+2}) + C(b_{n+2}-b_{n+3}) + \ldots$$

$$\ldots + C(b_{m-1}-b_m) + Cb_{n+1} + Cb_m$$

$$= 2Cb_{n+1} \to 0 \quad \text{as} \quad n \to \infty.$$

So $\{t_n\}$ is a Cauchy sequence in the complex plane and so $\{t_n\}$ converges, i.e. $\sum a_n b_n$ converges.

275. A USE OF DIRICHLET'S TEST

EXAMPLE. Discuss the convergence of $\sum z^n/n$ on its circle of convergence.

Solution. The radius of convergence is 1. So consider a general point $z = e^{i\alpha}$ $(0 \leq \alpha < 2\pi)$ on the circle of convergence. The series is then $\sum e^{in\alpha}/n$ and we can take $a_n = e^{in\alpha}$ and $b_n = 1/n$ in Dirichlet's test.

Certainly $\{b_n\}$ decreases to zero. Also

$$\left| \sum_{k=1}^n e^{ik\alpha} \right| = \left| \frac{e^{i\alpha}(1 - e^{in\alpha})}{(1 - e^{i\alpha})} \right| \qquad (\alpha \neq 0)$$

$$\leq \frac{2}{|-e^{\frac{1}{2}i\alpha}(e^{\frac{1}{2}i\alpha} - e^{-\frac{1}{2}i\alpha})|} \qquad (\alpha \neq 0)$$

(using the triangle inequality and the fact that $|e^{i\beta}| = 1$ for all $\beta\epsilon R$)

$$= \frac{1}{|\sin \frac{1}{2}\alpha|} \qquad (\alpha \neq 0).$$

So, provided that $\alpha \epsilon\ ]0,2\pi[$, Dirichlet's test tells us that $\sum e^{in\alpha}/n$ converges. When $\alpha = 0$ the series is $\sum 1/n$, which diverges.

So the series converges at all points of its circle of convergence except for the point $z = 1$.

<u>Notice</u>: (i) Since $e^{in\alpha} = \cos n\alpha + i \sin n\alpha$, the last example shows that the series

$$\sum \left(\frac{\cos n\alpha}{n} + i \frac{\sin n\alpha}{n} \right)$$

converges provided that $\alpha \neq 2k\pi$ $(k \in N)$. So, on splitting it into real and imaginary parts, we can rapidly deduce that

$\sum \dfrac{\cos n\alpha}{n}$ converges provided that $\alpha \neq 2k\pi$ $(k \in N)$

and $\sum \dfrac{\sin n\alpha}{n}$ converges for all $\alpha \in R$.

(ii) The above method will give a similar result for series like $\sum z^n/\sqrt{n}$ and $\sum z^n/\log n$.

(iii) The series $\sum z^n/n^2$ does <u>not</u> require the above treatment since when $|z| = 1$, $|z^n/n^2| \leq 1/n^2$ and so $\sum z^n/n^2$ is absolutely convergent at all points of its circle of convergence.

276. GAUSS'S TEST

This is a more delicate form of the ratio test. For a series $\sum a_n$ of positive terms the ratio test fails to reach a conclusion when $a_{n+1}/a_n \to 1$ as $n \to \infty$. Gauss's test often resolves this situation.

<u>RESULT</u>. (Gauss's test) Let $a_n > 0$ for all $n \in N$. Suppose that

$$\frac{a_{n+1}}{a_n} = 1 - \frac{k}{n} + O\left(\frac{1}{n^2}\right) \quad \text{as} \quad n \to \infty.$$

Then $\sum a_n$ converges if $k > 1$ and diverges if $k \leq 1$.

<u>Proof</u>. First take the case where $k > 0$. The result of §272 then applies to give the existence of a positive constant B such that $a_n \sim B/n^k$. So if we take $b_n = 1/n^k$ in the comparison test, we see that $a_n/b_n \to B$ as $n \to \infty$. So, by comparison with $\sum 1/n^k$ we conclude that $\sum a_n$ converges if $k > 1$ and diverges if $0 < k \leq 1$.

For the cases of $k < 0$ and $k = 0$ the results of §152 and §268 respectively give divergence for $\sum a_n$.

EXAMPLE.   Discuss the convergence of the series

$$\sum \frac{(2n)!}{(n!)^2}\left(\frac{1}{4}\right)^n .$$

Solution.   Here

$$\frac{a_{n+1}}{a_n} = \frac{(2n+2)(2n+1)}{4(n+1)(n+1)} = \frac{(n+\frac{1}{2})}{(n+1)} = 1 - \frac{1}{2n} + O\left(\frac{1}{n^2}\right).$$

Since $\frac{1}{2} < 1$, the series diverges by Gauss's test.

EXAMPLE.   Discuss the convergence of $\sum \frac{(2n)!}{(n!)^2} z^n$ at all points of its circle of convergence.   [Hint: The radius of convergence is 1/4.   The previous example proves divergence at $z = 1/4$.   At other points let $a_n = e^{in\alpha}$ and $b_n = \frac{(2n)!}{(n!)^2}(1/4)^n$ in Dirichlet's test. Use §272 to see that $\{b_n\}$ decreases to zero as $n \to \infty$.]

EXAMPLE.   Discuss the convergence of

$$\sum \frac{2.7.12...(5n-3)}{6.11.16...(5n+1)} z^n$$

at all points of the complex plane.

EXAMPLES 12

1.   Use Stirling's approximation to show that

$$\binom{2n}{n} \sim \frac{2^{2n}}{\sqrt{(\pi n)}} \quad \text{as} \quad n \to \infty.$$

2.   Find the radius of convergence R of $\sum \frac{n^n}{n!} z^n$.
Use Stirling's approximation to show that the series diverges for $z = R$.

3.   Let $\{f(n)\}$ be the sequence defined by

$$f(n) = \frac{1.4.7...(3n-2)}{3.6.9...(3n)} n^{2/3}.$$

Show that $\{f(n)\}$ tends to a non-zero limit as $n \to \infty$.

4.   Let $f(n) = \frac{3.6.11...(n^2+2)}{2.5.10...(n^2+1)}$ $(n \epsilon N)$.   Decide

by considering $f(n+1)/f(n)$ whether $\{f(n)\}$ remains bounded as $n \to \infty$.

5. Show that $\sum \dfrac{z^n}{\log n}$ has radius of convergence equal to 1. Use Dirichlet's test to discuss the convergence of this series at all points of its circle of convergence. Hence discuss the convergence of

$\sum \dfrac{\cos n\theta}{\log n}$ and $\sum \dfrac{\sin n\theta}{\log n}$ for all $\theta \, \varepsilon \, R$.

6. In §275 it is proved that $\sum \dfrac{\cos n\alpha}{n}$ converges for all $\alpha \neq 2k\pi$ $(k\varepsilon N)$. Deduce using a trigonometric identity that $\sum \dfrac{\cos^2 n\theta}{n}$ diverges for all $\theta \, \varepsilon \, R$. Deduce from this that the series $\sum \dfrac{|\cos n\theta|}{n}$ diverges for all $\theta \, \varepsilon \, R$.

7. In §275 it is proved that $\sum \dfrac{\sin n\alpha}{n}$ converges for all $\alpha \, \varepsilon \, R$. In contrast to this give and justify a particular value of α such that $\sum \dfrac{\sin n^2\alpha}{n}$ diverges.

8. The power series $\sum z^{n!}/n$ has radius of convergence equal to 1. Show that this series diverges at every point of the form $e^{2\pi\alpha i}$ where α is rational.

9. Discuss the convergence for all $z\varepsilon C$ of the series $\sum \dfrac{1.3.5 \ldots (2n-1)}{4.6.8 \ldots (2n+2)} z^n$.

10. (Behaviour of the binomial series on its circle of convergence)

The binomial series $\sum \begin{pmatrix} \alpha \\ n \end{pmatrix} z^n$, where $\alpha \, \varepsilon \, R-\{0,1,2,\ldots\}$, has radius of convergence equal to 1. Show that this series is absolutely convergent at all points z with $|z| = 1$ in the case when $\alpha > 0$ and that it is divergent at all points with $|z| = 1$ in the case when $\alpha \leq -1$. Use Dirichlet's test to discuss the convergence of the series at all points with $|z| = 1$ in the case when $-1 < \alpha < 0$.

EXAMPLES 13 - MISCELLANEOUS

(It is not intended that these be done in the order given.)

1.   (Compare Ex.3.18.)   A calculator with a square root key is to be used to find the seventh root of the positive number a.   Devise an algorithm to do this. Investigate the possibility of an algorithm to find the fifth root of a, and in general an algorithm for $a^{1/p}$ where p is an odd prime.

2.   A bounded continuous function on $[0,\infty[$ attains neither its supremum nor its infimum.   Show that it oscillates boundedly as $x \to \infty$.

3. $\{a_n\}$ is a positive sequence and $a_{n+1}/a_n \to L$ as $n \to \infty$.   By using a multiplicative collapse argument or otherwise, prove that $a_n^{1/n} \to L$ as $n \to \infty$.   (This is a standard result.)

4.   Prove that a periodic continuous function on R is uniformly continuous on R.

5.   Prove that if $f(x) \leq 0$ and $f'(x) + f(x) \geq 0$ for all $x \in R$ then $f(x) \to 0$ as $x \to \infty$.

6.   Notice that, since by §275 the series $\sum \dfrac{\sin nx}{n}$ converges for all $x \in R$, we can define a function

$$\phi: R \to R \quad \text{by} \quad \phi(x) = \sum_{n=1}^{\infty} \frac{\sin nx}{n} . \quad \text{The result}$$

of §198 on differentiation of <u>power</u> series may be better appreciated if attempts are made to differentiate the series for ϕ.   Notice that in general the term by term derivative of this series does not even converge.

7.   Prove that if $a_n \geq 0$ (n∈N) and

$$n^2(a_{n+1} + a_{n+2} + \ldots + a_{2n}) \to 0 \quad \text{as} \quad n \to \infty$$

then $\sum a_n$ converges.

8.   Prove that $\displaystyle\sum_{k=1}^{n} \frac{\log k}{k} \sim \frac{1}{2}(\log n)^2$ as $n \to \infty$.

9.   Apply the mean-value theorem to the function log x to show that, for every n∈N,

$$\frac{1}{n+1} \leq \log(n+1) - \log n \leq \frac{1}{n} . \qquad \ldots (1)$$

Sequences $\{f(n)\}$ and $\{g(n)\}$ $(n \varepsilon N)$ are defined by

$$f(n) = 1 + \frac{1}{2} + \frac{1}{3} + \ldots + \frac{1}{n} - \log n,$$

$$g(n) = 1 + \frac{1}{2} + \frac{1}{3} + \ldots + \frac{1}{n} - \log(n+1).$$

Use (1) to show that $\{f(n)\}$ is decreasing and that $\{g(n)\}$ is increasing. Deduce that $\{f(n)\}$ and $\{g(n)\}$ tend to a common limit γ as $n \to \infty$, and further that

$$\gamma + \log n \leq 1 + \frac{1}{2} + \frac{1}{3} + \ldots + \frac{1}{n} \leq \gamma + \log(n+1). \quad \ldots (2)$$

Using (1) and (2), prove that, for all $n \varepsilon N$,

$$\frac{1}{n+1} + \frac{1}{n+2} + \ldots + \frac{1}{2n} \geq \log 2 - \frac{1}{n}.$$

10. By considering its derivative show that the function $x/(\sin x)$ is increasing on $]0, \frac{1}{2}\pi[$ and deduce that, for all $x \ ]0, \frac{1}{2}\pi[$,

$$\frac{2x}{\pi} < \sin x < x.$$

[The result of Ex.10.51 may help in this.]

11. Use the result of §180 to show that

$$\lim_{n \to \infty} \frac{1}{4n+1} + \frac{1}{4n+5} + \ldots + \frac{1}{8n-3} = \frac{1}{4} \log 2.$$

Gregory's series (§220) is rearranged as

$$1 + \frac{1}{5} - \frac{1}{3} + \frac{1}{9} + \frac{1}{13} - \frac{1}{7} + \frac{1}{17} + \frac{1}{21} - \frac{1}{11} + \ldots .$$

Prove that this series converges and use the above limit to find its sum.

12. Using the series for $\log(1 + x)$ it is possible to calculate easily

$$\log \frac{9}{10}, \quad \log \frac{11}{10}, \quad \log \frac{81}{80}, \quad \log \frac{121}{120}, \quad \log \frac{2401}{2400}.$$

Show how these five values can be used to find the logarithms of 2, 3, 4, 5, 6, 7, 8, 9, 10, 11.

13. Show that $\left| \left(1 + \frac{i}{\sqrt{n}} \right)^n - \left(1 - \frac{i}{\sqrt{n}} \right)^n \right|$ stays bounded as $n \to \infty$.

14. Prove that $2 \tan^{-1} \frac{7}{17} + \tan^{-1} \frac{1}{239} = \frac{1}{4}\pi$.

Does the number 239 appear both here and in Machin's

H

formula (§220) entirely by chance?   Given the Pythagorean triple (20,21,29) construct another relation of the same type expressing $\pi/4$ as a linear combination of inverse tangents of rational numbers.

15.   Find explicitly the sum of the series

$$\frac{1}{2} - \frac{2}{3} + \frac{1}{4} + \frac{1}{6} - \frac{2}{7} + \frac{1}{8} + \frac{1}{10} - \frac{2}{11} + \frac{1}{12} + \cdots .$$

16.   Show that

$$(\tfrac{1}{2}) + \frac{1}{5}(\tfrac{1}{2})^5 + \frac{1}{9}(\tfrac{1}{2})^9 + \cdots = \tfrac{1}{2}\tan^{-1}\tfrac{1}{2} + \frac{1}{4}\log 3 .$$

17.   A sequence is defined by $f(n) = n^n/n!$.   Show that $f(n+1)/f(n) \to e$ as $n \to \infty$, and use the result of Ex.3 above to deduce that

$$(n!)^{1/n} \sim n/e \quad \text{as} \quad n \to \infty .$$

18.   Let $0 < a < 1$.   Define $\{f(n)\}$ by $f(n) = n^{\log n}a^n$ $(n \epsilon N)$.   Prove that $f(n) \to 0$ as $n \to \infty$.

19.   Let $0 < a < 1$.   Evaluate $\lim\limits_{n \to \infty} n^n a^{n!}$.

20.   Prove that $\sum \dfrac{1}{n^z}$ is absolutely convergent if Re $z > 1$.   Prove, using Ex.10.35, that $\sum \dfrac{1}{n^{1+i}}$ is divergent.

21.   Show that, for $\alpha > 0$ and $\beta > 0$,

$$\sum_{r=1}^{n} r^\alpha(n-r+1)^\beta \sim B(\alpha+1,\beta+1)n^{\alpha+\beta+1} \quad \text{as} \quad n \to \infty .$$

22.   If you were asked to give a short talk about the relative advantages and disadvantages of developing the exponential function (on R or on C) by the two methods mentioned in §208, what points would you make?

APPENDIX

---

277. CAUCHY'S INEQUALITY AND THE GENERAL TRIANGLE INEQUALITY

In R^3 the ideas of the <u>length</u> $\|x\|$ of a vector x and the <u>scalar product</u> $x \cdot y$ of two vectors x and y are well-known. These ideas can be generalised to R^n ($n \geq 3$) and indeed to a general Hilbert space. The following is a basic result in Hilbert space theory. We include it because it provides two useful inequalities in R^n.

<u>RESULT</u>. Let H be a real or complex Hilbert space. Let $\|x\|$ denote the norm (i.e. length) of the vector x and let (x, y) denote the scalar product of the vectors x and y. Then, for all $x, y \, \varepsilon \, H$,

(1) $|(x, y)| \leq \|x\| \|y\|$ [<u>Cauchy's inequality</u>]

(2) $\|x + y\| \leq \|x\| + \|y\|$ [<u>Triangle inequality</u>]

<u>Proof</u>. We prove the result in the complex case. When $x = y = 0$ the result is trivial. So we can assume without loss of generality that $x \neq 0$. Then, for every $\alpha \varepsilon C$,

$$(\alpha x + y, \alpha x + y) \geq 0,$$

i.e. $|\alpha|^2 (x, x) + \alpha(x, y) + \bar{\alpha}(y, x) + (y, y) \geq 0$. This is true for <u>every</u> $\alpha \varepsilon C$ so that we are at liberty to choose $\alpha = -(y, x)/(x, x)$ in this relation. It follows that

$$- \frac{|(x, y)|^2}{(x, x)} + (y, y) \geq 0.$$

Hence $|(x, y)|^2 \leq (x, x)(y, y)$, which gives (1).

For (2), notice that

$$\|x+y\|^2 = (x+y, x+y) = (x, x) + (x, y) + (y, x) + (y, y)$$

$$= \|x\|^2 + 2 \, \text{Re}(x, y) + \|y\|^2$$

$$\leq \|x\|^2 + 2|(x, y)| + \|y\|^2$$

$$\leq \; \|x\|^2 + 2\|x\| \cdot \|y\| + \|y\|^2 \qquad \text{(by (1))}$$

$$= \; (\|x\| + \|y\|)^2 .$$

This proves (2).

In R^n, the <u>scalar product</u> of the vectors
$x = (x_1, x_2, \ldots , x_n)$ and $y = (y_1, y_2, \ldots , y_n)$ is
given by

$$(x , y) = x_1 y_1 + x_2 y_2 + \ldots + x_n y_n$$

and the <u>length</u> of the vector x is given by

$$\|x\| = (x_1^2 + x_2^2 + \ldots + x_n^2)^{\frac{1}{2}} .$$

In the setting of the Hilbert space R^n the above
inequalities become the following.

(1)   (<u>Cauchy's inequality for real numbers</u>)

Let $x_1, x_2, \ldots , x_n, y_1, y_2, \ldots , y_n$ be real numbers.

Then
$$\left| \sum_{i=1}^{n} x_i y_i \right| \leq \left(\sum_{i=1}^{n} x_i^2 \right)^{\frac{1}{2}} \left(\sum_{i=1}^{n} y_i^2 \right)^{\frac{1}{2}} .$$

(2)   (<u>The general triangle inequality</u>)

Let $x_1, x_2, \ldots , x_n, y_1, y_2, \ldots , y_n$ be real numbers.

Then
$$\left(\sum_{i=1}^{n} (x_i + y_i)^2 \right)^{\frac{1}{2}} \leq \left(\sum_{i=1}^{n} x_i^2 \right)^{\frac{1}{2}} + \left(\sum_{i=1}^{n} y_i^2 \right)^{\frac{1}{2}} .$$

These are useful in certain situations, e.g. Ex.6.15.

SOME BOOKS

Some of the following deal with subjects beyond the scope of this book. A date does not necessarily refer to the most recent edition.

S.K.BERBERIAN, Introduction to Hilbert Space, Oxford, 1961.

B.R.GELBAUM & J.M.H.OLMSTED, Counterexamples in Analysis, Holden-Day, 1964.

E.HILLE, Analytic Function Theory, Volume 1, Blaisdell, 1965.

M.MANSFIELD, Introduction to Topology, Van Nostrand, 1963.

R.A.RANKIN, An Introduction to Mathematical Analysis, Pergamon, 1963.

W.RUDIN, Real and Complex Analysis, McGraw-Hill, 1974.

W.A.SUTHERLAND, Introduction to Metric and Topological Spaces, Oxford, 1975.

COMMENTS ON AND ANSWERS TO THE EXAMPLES

(Notice that the references are to sections <u>not</u> pages.
Notice also that * after a section number refers to an
example within that section.)

EXAMPLES 1 (PAGES 20-23)

1.　　You will need the meaning of $c < 0$ from §4.
Then use $-c(a - b) > 0$.

3.　　For (i) use $(x^2 - y^2) = (x - y)(x + y)$.　　For (ii)
and (iii) see §5.

4.　　See §9*.

5.　　See §8* or §11(i).

6.　　Write $a + b = a - (-b)$ and use §9.

7.　　Suppose $a > 2$.　　Then notice that $0 < 2/a < 1$.
So take $b = 2/a$.　　Then from the given hypothesis
$ab < 2$.　　This gives a contradiction.

8.　　See §11(i).

9.　　Similar to §11(ii). For $|1 - x| \geq 1 - |x| \geq 1/4$.

10.　　For $x \varepsilon [3,4]$ the top is at least 540, while the
bottom is at most 529.

11.　　Use triangle inequality on the LHS and use
$2/n^3 \leq 2/n$ etc.

12.　　(a)　Proceed as in §13(iii) to find $K = 8$ will
do.　　(b)　Proceed as in §13(iv) to find $K = 1/4$ will do.
(c)　Take the LHS over a common denominator first.
$K = 16$ will do.

13.　　Proceed as in §13(iii) for the RHS and as in
§13(iv) for the LHS.

15.　　Proceed as in §20*.

16.　　Solution given in §36.

17.　　Let $M > 0$ be an upper bound for A.　　Then
$1/M$ is a lower bound for B.　　To show $\inf B = 1/(\sup A)$
proceed as in §20*.　　For the last part, notice that
if B had an upper bound $K > 0$, then $1/K$ would be a
<u>positive</u> lower bound for A.　　This would contradict the
fact that $\inf A = 0$.

18.　　A = all rationals in]0,1[,　B = all irrationals
in]0,1[.

19.　　(a)　Let $|f(x)| \leq M$ and $|g(x)| \leq N$ for all $x \varepsilon R$.
Then $|f(x)g(x)| \leq MN$ for all $x \varepsilon R$.

(b)　Notice first that g is <u>positive</u> so that $1/g$
is bounded below by 0.　　Also since $g(x) > \alpha$ for all
x R, $1/g$ is bounded above by $1/\alpha$.

(c) $g(x) = |x|$ $(x \neq 0)$, $g(x) = 1$ $(x = 0)$.

20. Proceed as in §38 working on the interval $]3,4[$ and estimating to retain the factor $(x - 3)$ only.

21. See §38. Work on $]6,7[$ or $]4,5[$.

22. Let $f(x) = 0$ (x rational), $f(x) = x$ (otherwise) and let $g(x) = x$ (x rational), $g(x) = 0$ (otherwise).

23. Use $|f(x) \pm g(x)| \leq |f(x)| + |g(x)|$ for the first part. For the second part, suppose that $(h + k)$ is bounded while h is unbounded and k is bounded. Then, by the first part, $(h + k) - k$ is bounded. Contradiction.

24. Show that $1/g$ is bounded and then use a contradiction argument similar to Ex.23.

26. See §28.

27. Notice that the functions are positive. Also inf f $\leq$ f(x) for all x$\in$D. Multiply both sides by g(x) to get the first part. If inf f = 0 the second part is trivial. If inf f $\neq$ 0 the first part gives $g(x) \leq \sup(fg)/(\inf f)$ $\forall x \in D$; in this the RHS is a <u>constant</u>, and so is an upper bound for g. So, $\sup g \leq \sup(fg)/(\inf f)$. Hence result.

28. $1/(x^2+1)$.

30. Take the LHS as $\frac{1}{2}(x + \frac{\alpha}{x})$.

31. For the second part notice that there are 2n terms on the LHS and each exceeds $2/(2n+1)$.

32. Use a contradiction argument. Suppose that $\sqrt{2} = m/n$, where $m, n \in N$ and m, n have no common factor. Then $m^2 = 2n^2$. Since the RHS has the factor 2 it follows that 2 divides m. So write m = 2p. But then it follows that 2 divides n also. Contradiction.

33. Either $1 > 0$ or not. Follow these possibilities up.

EXAMPLES 2   (PAGES 41-45)

1. Proceed as in §44 and §46 respectively.

2. Proceed as in §53. The limits are 3, $\frac{3}{4}$ and 0 respectively.

3. Similar to §54*.

4. Similar to §54* except that the limit on each side is 1.

5. Use §5 and §51.

6. (a) $(-1)^n$, (b) $(-1)^n$, (c) $(-1)^n$, (d) $(-1)^n$, (e) $(-1)^n/n$, (f) $1/\sqrt{n}$. To see that the answer to (f) is satisfactory, notice that we have that

$$\frac{1}{\sqrt{n}}(a_{n+1}+a_{n+2}+ \cdots +a_{2n}) > \frac{1}{\sqrt{n}}\left(n \cdot \frac{1}{\sqrt{(2n)}}\right) = \frac{1}{\sqrt{2}} \cdot$$

7. $f(n+1) < f(n)$ for all $n > X$.   So the sequence is eventually decreasing and bounded below by 0.   So $f(n) \to L$.   But $f(n+1) < \alpha f(n)$.   So $L \leq \alpha L$. This gives a contradiction unless $L = 0$.   (Compare §65*.)   For the example take $f(n) = (n+1)/n$.

8. $\frac{f(n+1)}{f(n)} = \frac{3(2n+2)}{4(2n+1)} \to \frac{3}{4}$ as $n \to \infty$.   Then proceed as in §65*.

9.   Similar to §57*.   Notice that
$$a_{n+1} = \frac{1}{2n+3} + \cdots + \frac{1}{4n-1} + \frac{1}{4n+1} + \frac{1}{4n+3} \cdot$$

10.   Use §48(i) with $\varepsilon = 1/4$.   Then use $(1/4)^n < (f(n))^n < (3/4)^n$ with the sandwich result and §62.   In the three cases in the generalisation, $(f(n))^n$ tends to zero, tends to infinity, oscillates unboundedly.

11. Notice that $na > 2$ for all $n > 2/a$.   It follows that $(na)^n > 2^n$ for all such   n.   So $(na)^n \to \infty$.

12.   The limits are $0, \infty, 0, \infty, 0$ respectively. For (a) split it into the sum of two terms and use §63. For (b) invert (a).   For (c) copy §63*.   For (d) write 2^{2n} as 4^n, estimate and use §63.   For (e) multiply above and below by $\sqrt{(n+1)} + \sqrt{n}$.

13.   (a) $a_n = 2n$, $b_n = -n$.   (b) $a_n = b_n = (-1)^n$.

(c) $a_n = (-1)^n$, $b_n = -(-1)^n$.   (d) $a_n = 1/n$, $b_n = 1/n^2$.

(e) $a_n = n^2 + n$, $b_n = n^2$.

14.   (a) $a_n = 2n$ (n even), $a_n = n$ (n odd).

(b) $a_n = 3/n$ (n even), $a_n = 1/n$ (n odd).   (c) $a_n = n^2$.

(d) $a_n = (-1)^n/n$.   (e) Take $\{1, 1, -\frac{1}{2}, -\frac{1}{2}, \frac{1}{3}, \frac{1}{3}, -\frac{1}{4}, \dots\}$ as the sequence.   (f) $a_n = n^{-1/4}$.   (g) $a_n = 1/2^n$.

15.   Choose $\varepsilon > 0$.   Then $|f(n)| \leq M$ for all $n \varepsilon N$.

16.   First notice that $\{f(n)g(n)\}$ is bounded. So either it tends to a limit L or it oscillates boundedly.   If it tends to a limit then so also does $f(n)g(n) \cdot \frac{1}{g(n)}$ by §51.   Contradiction.

17.   Choose $\varepsilon = 1/4$, and use §48(iii) to get a contradiction.

18.   Let $M > 0$ be the supremum.   Take $\varepsilon = \frac{1}{2}M$. Then from the definition of a limit $a_n < \frac{1}{2}M$ for all $n \geq X$.   There are only finitely many integers less

than X and k must be among them.

19.   Treat L = 0 and L > 0 separately.

20.   Treat L = 0 and L > 0 separately.   For
L > 0, use $a_n^3 - L^3 = (a_n - L)(a_n^2 + La_n + L^2)$.

21.   Similar to §64(i).   For (b), notice that if
$a_1 < 0$ then $a_2 > 2$.

22.   Similar to §64(i).

23.   For the "deduce" write $(a_{n+1}-4)=(a_n-4)/(a_{n+1}+4)$.
Then the last part is similar to §64(ii).

24.   Similar to §64(ii).

25.   The sequence is bounded.   So either it tends to
a limit or it oscillates boundedly.   Suppose it tends to
a limit.   Use §48(iii) with $\varepsilon = (p-q)/4$, where p and
q are the supremum and infimum thereby getting a
contradiction.

26.   Suppose not.   Then $\neg(\forall X, \exists N > X$ such that
$|a_{N+1} - L| < |a_N - L|)$.   Go for a contradiction.

27.   Compare §65.

28.   Suppose $\forall n \varepsilon N$, $a_n \neq 2$ and $a_n \neq 7$ and $a_n \rightarrow L$.
Then $L = L^2 - 8L + 14$. So L = 2 or 7. Suppose L = 7.
So $|a_n - 7| < 1$ $\forall n \geq N$, say. Now show $|a_{N+1} - 7| > 5|a_N - 7|$,
$|a_{N+2} - 7| > 5^2|a_N - 7|$, etc. So $|a_k - 7| \rightarrow \infty$ as $k \rightarrow \infty$.

29.   A standard result.

30.   Suppose $\{f(n)g(n)\}$ is bounded.   What are then
the implications for g of the fact that $f(n) \rightarrow \infty$?

31.   Start as in §65. $a_{n+1}/a_n \rightarrow \infty$ as $n \rightarrow \infty$.
Prove from this that $a_n \rightarrow \infty$ as $n \rightarrow \infty$.

32.   A standard result.

EXAMPLES 3   (PAGES 57-61)

1. $f(0+) = 3$, $f(0-) = -3$.   No, by §72.

2. $f(1+) = 1$, $f(1-) = 0$.   No, by §72.   For

the limit as $x \rightarrow \infty$, use $\dfrac{x-1}{x} \leq \dfrac{[x]}{x} \leq \dfrac{x}{x}$ and a sandwich
result.

3.   (a) $k \geq 3$, (b) $k \leq 1$, (c) k = 2.

4.   Similar to §38.   The only real difference is
that you are to produce δ and not a value of x as in
§38.

5.   As Ex.4 above.

6.   Similar to §85.

7.   (a) R, (b) R - Z, (c) R - Z.

8.   Similar to §85.   g(x) = -1 (x rational),
g(x) = 1 (x irrational).

9.   Similar to §85.

10.   Use the intermediate value theorem (§87) on each
interval.

11.   Use §87 together with §12.

12.   (i) f(x) = x + 1.   (ii) f(x) = x^2.

13.   f $\circ$ f is continuous and maps [a,b] into [a,b].
So apply §88 to find k such that f(f(k)) = k.   Then
let p = f(k), q = k.   For the "deduce" notice that
(p,q) and (q,p) both lie on the graph of f and apply
§87 to g, where g(x) = f(x) - x.

14.   §87 is needed.

15.   §87.

16.   Suppose not and use §66 together with the fact
that f is continuous.

17.   All the limits are 1.

18.   (i) Induction.   (ii) Take L = $\alpha^{1/3}$ in the
first part.

19.   f(x) = -1/x.

20.   Compare Ex.3 above and split your argument into
three cases according as the degree of the numerator is
greater than, less than or equal to that of the
denominator.

21.   (f+g) is certainly bounded.   So either it tends
to a limit or it oscillates.   Suppose it tends to a
limit.   Then so also does (f+g)-g by heredity.
Contradiction.

22.   R - Z.

23.   g = ½(g$_1$ + g$_2$).   For uniqueness suppose that
g = e$_1$ + o$_1$ = e$_2$ + o$_2$.   Then e$_1$ - e$_2$ = o$_2$ - o$_1$.   The LHS
is even, the RHS is odd.   So e$_1$ - e$_2$ is both even and
odd, and the only such function is the zero function.

24.   Compare §51.

25.   Copy §57.

26.   Suppose that f(x) → ∞.   Let K = f(0) + 1.
Then f(x) > K for all x > X, say.   Take x = np,
where p is the period and n∈N is large enough to make
np > X.   Then f(np) = f(0) by periodicity.
Contradiction.
    Yes for the last part.

27.   For the second part notice that f(1) = (f(½))^2 ,
which shows that f(1) $\geq$ 0.   Then prove that

f(1/2^n) = (f(1))$^{1/2^n}$.   Now use §92 and the continuity
of f.

28.   (i) $\Rightarrow$ (ii) is easy.   To show (ii) $\Rightarrow$ (i), suppose that (ii) holds but f is discontinuous at a. So $\neg\,(\,\forall\,\epsilon > 0,\;\exists\,\delta > 0$ such that

$$\forall\,x\,\epsilon\,]a-\delta,a+\delta[,\quad |f(x) - f(a)| < \epsilon),$$

i.e. $\exists\,\epsilon > 0$ such that $\forall\,\delta > 0$

$$\exists\,x\,\epsilon\,]a-\delta,a+\delta[\text{ such that }|f(x) - f(a)| \geq \epsilon.$$

Using this last relation, choose successively $\delta = 1, \frac{1}{2}, \frac{1}{3}, \dots , \frac{1}{n}$ thereby locating points $a_1, a_2, a_3,$ $\dots , a_n$ with $a_n\,\epsilon\,]a-\frac{1}{n},a+\frac{1}{n}[$ and $|f(a_n) - f(a)| \geq \epsilon.$ Then $a_n \to a$ but $f(a_n) \not\to f(a)$.   This contradicts the assumption of (ii).

29.   For the last part notice that, for <u>every</u> non-negative function g, $f = (f^+ + g) - (f^- + g)$.

EXAMPLES 4   (PAGES 78-81)

1.   (a) M1 fails because $d(2,2) \neq 0$.   (b) M3 fails because $d(0,2) = 4$ while $d(0,1) + d(1,2) = 2$. (c) M1 fails because $d(-1,1) = 0$.   (d) M1 fails because $d(3,3) = 1 \neq 0$.

2.   M1 and M2 are clear.   For M3 use the ordinary triangle inequality for real numbers.   The ball is the square with vertices at (1,0), (0,1), (-1,0) and (0,-1). Equivalence can be checked with the criterion of §99.

3.   For the second part use $x^2 - y^2 = (x + y)(x - y)$. For the "deduce", notice that if $x_n \to a$ with respect to d_1, then $d_1(x_n,a) \to 0$.   But then $d_2(x_n,a) \leq 10\,d_1(x_n,a)$.   So $d_2(x_n,a) \to 0$.   So $x_n \to a$ with respect to d_2.   So d_1-convergent sequences are d_2-convergent.   Use the other half of the inequality to show that d_2-convergent sequences are d_1-convergent.

4.   The given sequence has no limit in S.   So S is not complete.

6.   This is similar to §101.   N(0,1) is a disc. N(1,1) and N(1+i,1) are both line segments.   Of the sequences only (a) is convergent in the given metric. This metric is not equivalent to the usual metric, e.g. because we cannot put an open disc inside the line segments.

7.   Use both $d(x,z) \leq d(x,y) + d(y,z)$ and $d(y,z) \leq d(y,x) + d(x,z)$, rather as in §9.

8.   Only (c), (e) and (g) are closed.

9.   (a) [1,3], (b) [1,3], (c) R, (d) the point 0
only, (e) no limit point exists.

10.   The sets A_n (nεN), where A_n =]1/n,2[.

11.   Let σ = sup f.   Then from the fact that
f(x) $\rightarrow$ 0 as x $\rightarrow$ $\pm\infty$, we can find an interval [-N,N]
such that f(x) < $\frac{1}{2}\sigma$ for all x ε]-∞,-N[ ]N,∞[.
Then apply §107(ii) on [-N,N].   For the example take
f(x) = $1/(x^2+1)$.

13. $|f(x) - f(y)| \leq 21|x - y|$ on [-4,4].

14.   (a) says that f is uniformly continuous on R,
while (b) says that f is continuous on R.

15.   By the triangle inequality $|f(2\delta) - f(0)| \leq$
$|f(2\delta) - f(\delta)| + |f(\delta) - f(0)| \leq 2$.   Similarly
$|f(n\delta) - f(0)| \leq$ n.   So $|f(n\delta)| \leq |f(n\delta) - f(0)| + |f(0)|$
$\leq$ n + $|f(0)|$.
    For a general point x, choose an integer n with
n$\delta \leq$ x $\leq$ (n+1)δ.   Hence find A and B.

16.   For the example, take f(x) = g(x) = x.

17.   (a) Yes.   (b) No.   (c) Yes.   (d) No.
(e) Yes: see §136.   (f) No: the oscillations speed up.
(g) No: as x $\rightarrow$ ∞, the function behaves like x^2 and so
does not satisfy a criterion like that of Ex.15 above.
(h) Yes: the idea of §136 applies.

18.   Proceed by contradiction.   Choose points
x = $(2k\pi)^{\frac{1}{2}}$, y = $((2k+1)\pi)^{\frac{1}{2}}$, where k$\varepsilon$N.

19.   Suppose that (i) is false.   Then negating (i)
gives a particular δ such that f(x) $\leq$ $\frac{1}{2}$ on [0,δ].
Now apply the result of §107 on [δ,1] keeping in mind that
f does not take the value 1.   So there is an upper
bound k < 1 for f on [δ,1] and there is an upper
bound $\frac{1}{2}$ for f on [0,δ].   This is not consistent
with the fact that sup f = 1.

21.   Let a be any point of S.   The sets N(a,1),
N(a,2), N(a,3), ... form an open covering of S.   But
S is compact.   So there is a finite subcovering and
from the sets of it we select the one with the largest
radius, say N(a,p).   Then any two points in N(a,p)
are less than 2p apart.

24.   Choose ε > 0.   First use the fact that
f(x) $\rightarrow$ 0 as x $\rightarrow$ $\pm\infty$ to find N such that
$|f(x)|$ < $\frac{1}{4}\varepsilon$ for all $|x| \geq$ N.   So $|f(x) - f(y)|$ < $\frac{1}{2}\varepsilon$ < ε
for all x,y $\geq$ N and for all x,y $\leq$ -N.

    Then quote §111 to show f is uniformly continuous
on [-N,N].   This produces δ.

    Lastly consider the case of points x, y with
$|x - y|$ < δ and x ε [-N,N] and y ε [N,∞[.

25.   Look at [0,p], where p is the period of f.

EXAMPLES 5   (PAGES 95-98)

2.   Similar to §131.

3.   Similar to §132.   You may need §29 for the second equation.

4.   Let $f(x) = \sin x - x + x^3/6$.   Find $f(0)$ and $f'(x)$ as in Ex.2 above.   You will need to know the result from §131 that $\cos x - 1 - \tfrac{1}{2}x^2 > 0$ for $x \in]0, \tfrac{1}{2}\pi[$.

5.   Apply the intermediate value theorem on $]-\infty, a]$, $[a,b]$ and $[b,\infty[$ with suitable care at endpoints.   This will locate points $x < y < z$ with $f(x) = f(y) = f(z) = 0$. Then apply Rolle on $[x,y]$ and on $[y,z]$ to find d and e with $f'(d) = f'(e) = 0$.   Then apply Rolle to f' on $[d,e]$.   For the example, $f(x) = x^3 - x$.

6.   Notice that $f(a) = f(b) = 0$.   Apply Rolle.

8. $f(x) = -1/x$.

9.   Suppose $A > 0$.   Then notice that $f'(x) > \tfrac{1}{2}A$ for all $x \geq N$, say.   Then apply §133(i) to see that $f(x) \to \infty$.   Similarly if $A < 0$, $f(x) \to -\infty$.

10.   Think of f'' as the derivative of f'.   So if f'' is positive this means that f' is increasing. A result like that of §57 then applies to f'.

11.   Similar to the proof of Rolle's result (§128).

12.   Suppose not and without loss of generality suppose that $B > 0$.   Then from Ex.9, $g'(x) \to \infty$ as $x \to \infty$.   Then from §133 it follows that $g(x) \to \infty$. Contradiction.

13.   Similar to Example (ii) in §135.

14.   Similar to the example in §135.   A suitable algorithm takes a_1 aribtrarily in $[1,2]$ and then $a_{n+1} = f(a_n)$ $(n \in N)$, where $f(x) = 15(x^2+12)^{-1}$.   Show that $|f'(x)| \leq \dfrac{90}{144}$ for $x \in [0,3]$.

15.   It turns out that $|a_{n+1} - \alpha| > |a_n - \alpha|$, whenever $a_n \in]3,4[$.   This contrasts with what happens in Ex.(ii) of §135. $\{a_n\}$ finally oscillates between 0.017 and 59.983.   (This actually illustrates Ex.3.13, with $f(0.017) = 59.983$ and $f(59.983) = 0.017$.)

16. $|g'(x)| = \dfrac{30|x^2 + 4x - 31|}{(x^2 + 31)^2} < \dfrac{30.19}{35^2}$ for all $x \in [2,5]$.   The method is similar to that in §135.

17.   See Example (ii) of §135.   The root is 3.8297 as in Ex.16 above.

18. Use a contradiction argument. Suppose that $|g(a) - g(b)| < \varepsilon$ whenever $|a - b| < \delta$. Then choose N such that $g'(x) > 4\varepsilon/\delta$ for all $x \geq N$. Then take $a = N$, $b = N + \frac{1}{2}\delta$. Apply the mean-value theorem to $|g(a) - g(b)|$ to get a contradiction.

19. For the unboundedness look at $f'(u)$, where $u = (2k\pi)^{1/4}$ with $k \varepsilon N$.

20. For the first part divide through by $|x - y|$. To deduce uniform continuity see §136.

21. $f(x) = \sin \sqrt{x}$.

22. $f(x) = (\sin (x^2))/x$.

23. Since $f'(x) > 0$ for all $x \varepsilon R$, f is strictly increasing on R. So $f(x) > f(0)$ for all $x > 0$. So $f'(x) \geq f(x) > f(0)$ for all $x > 0$, i.e. $f'(x) > f(0)$ for all $x > 0$. Now apply §133(i).

24. Think of g'' as the derivative of g'. Since g'' is positive, g' is strictly increasing. So g' either tends to a limit or to infinity. But then by §133(i), g' cannot tend to infinity, nor can it tend to a non-zero limit. So $g'(x) \to 0$ as $x \to \infty$.
   For the second part notice that g' must be negative on R (because it is increasing and tends to zero). So g is decreasing. Also g is bounded below by hypothesis. So $g(x)$ tends to a limit.

25. Go for a contradiction. If no such point d exists, $f(x) \geq f(a)$ for all $x > a$ and so

$$\frac{f(x) - f(a)}{x - a} \geq 0 \quad \text{for all} \quad x > a, \text{ and so } f'(a) \geq 0.$$

26. Use §107 to see that the infimum is <u>attained</u>. Use Ex.25 above to see that it is attained on the open interval. Use the method of proof of Rolle's theorem to see that $f'(c) = 0$. For the "deduce", suppose without loss of generality that $g'(a) < g'(b)$ and that $g'(a) < k < g'(b)$. Then apply the first part to the auxiliary function f defined by $f(x) = g(x) - kx$, to see that g' takes the value k.

EXAMPLES 6   (PAGES 118-122)

(Here and later we make the following abbreviations: C = convergent, D = divergent, AC = absolutely convergent, CC = conditionally convergent, CT = comparison test, RT = ratio test.)

1. (a) D: CT with $\sum 1/n$. (b) C: CT with $\sum 1/n^2$. (c) D: $a_n \not\to 0$. (d) D: RT. (e) C: RT. (f) C: CT with $b_n = n^2/3^n$. (g) C: RT. (h) D: $a_n \not\to 0$. (i) same answer as (h). (j) D: CT with $\sum 1/\sqrt{n}$.

(k) C: rationalise and use CT with $\sum 1/n^{3/2}$.

(l) C: RT, where $a_{n+1}/a_n = (n+1)^2/((2n+1)(2n+2)) \to 1/4$.

(m) D: RT, where $a_{n+1}/a_n \to 2$. (n) D: CT with $\sum 1/n$ and use §92. (o) C: RT, where $a_{n+1}/a_n \to 0$.

(p) C: RT, where $a_{n+1}/a_n = (8n+4)/(9n+15)$.

(q) D: method of §152. (r) C: RT, where $a_{n+1}/a_n \to 0$ by §63. (s) D: method of §152. (t) C: RT, where $a_{n+1}/a_n \to 0$. (Alternatively use Cauchy's root test from Ex.6.22.) (u) C: $\sqrt{n} > 2$ for all $n > 4$. So $1/n^{\sqrt{n}} < 1/n^2$ for all $n > 4$.

2. Use CT with $b_n = 1/n^{\beta-\alpha}$.

3. If $\sum (a_n+b_n)$ converges, then so also does $\sum ((a_n+b_n) - a_n)$ by §140(ii).

4. 4/11, 23/74. For the first part write the number as $36/10^2 + 36/10^4 + \ldots$ and sum this geometric series.

5. $s_n(1 - x) = 1 + x + x^2 + \ldots + x^{n-1} - nx^n = \dfrac{1-x^n}{1-x} - nx^n$. Now divide by $(1-x)$ and let $n \to \infty$.

6. Take over a common denominator and use CT.

7. Use a collapsing argument. The sum is 1/4.

8. (a) CC. (b) AC: $|(-1)^n \sin n| \leq 1$ for all n. (c) D: $a_n \not\to 0$. (d) AC: RT on $\sum |a_n|$.

9. (a) $-3 < x < 3$: D at $x = \pm 3$ since $a_n \not\to 0$. (b) all $x \in R$. (c) $-2 < x \leq 2$. (d) $x = 0$.

10. AC for $k < 1$; CC for $1 \leq k < 2$.

11. (a) The series exceeds $\frac{1}{5} + \frac{1}{10} + \frac{1}{15} + \frac{1}{20} + \ldots$. (b) It exceeds $\frac{1}{5} + \frac{1}{10} + \frac{1}{15} + \ldots$ since $1 > \frac{1}{2}$, $\frac{1}{3} > \frac{1}{4}$ etc. (c) It exceeds $1 - \frac{1}{2} - \frac{1}{3} + \frac{1}{4} + \frac{1}{9} + \frac{1}{14} + \ldots$.

12. $L + a^2 K$.

13. (a) $a_n = 1/n$. (b) $a_n = 1/n$. (c) $a_n = 1/n^2$, $b_n = 1/n$. (d) $\frac{1}{2^2} + 1 + \frac{1}{4^2} + \frac{1}{3^2} + \frac{1}{6^2} + \frac{1}{5^2} + \ldots$. (e) $a_n = b_n = (-1)^n/\sqrt{n}$. (f) $a_n = (-1)^n/\sqrt{n}$. (g) $a_n = 2^{-n}$. (h) $a_n = 1$ (n even), $a_n = 0$ (n odd), $b_n = 0$ (n even), $b_n = 1$ (n odd). (i) $a_n = 1/(n \log n)$.

13. (continued)   (j) $a_n = 1/(n \log n)$:   the sum given is less than $n \cdot \dfrac{1}{n \log n} \to 0$ as $n \to \infty$.   (k) $a_n = 1/\sqrt{n}$ if $n = k^4$ for some positive integer k, $a_n = 0$ for all other integers n.

14. $1 + \frac{1}{2} + \dots + \frac{1}{9} < 9$, $\frac{1}{11} + \frac{1}{22} + \dots + \frac{1}{99} < \frac{9}{10}$, $\frac{1}{101} + \frac{1}{111} + \dots + \frac{1}{999} < \frac{90}{100}$, etc.   Sum a geometric series.

15.   (a) $a_n \to 0$ as $n \to \infty$.   So there exists   N such that $0 \le a_n \le 1$ for all $n > N$.   But then $0 \le a_n^2 \le a_n$ for all $n > N$.   Use comparison.

16.   Both the real and imaginary parts converge by Leibniz.

17.   Use $|z|^2 = z \bar{z}$ and multiply out.

18.   Take the modulus to see that no convergence is possible if $|z| > 1$.   Also see that $|z| < 1$ gives convergence to zero.   If $|z| = 1$, notice by §164(ii) that any possible limit has modulus   1.   The only suitable values with $|z| = 1$ are 1, i, -1, -i: this can be proved by looking at consecutive terms as in §65.

19.   Suppose both converge.   Get a contradiction.

20.   Consider $f(x) - \sum\limits_{n=1}^{\infty} a_n \cos nx$.   Is this even, odd or both?   See also Ex.3.23.

21.   If the series converges and s_n denotes its nth partial sum, $(s_{2n} - s_n)$ should tend to zero as $n \to \infty$. Prove this is not so by estimating $(s_{2n} - s_n)$:   §57* will help in this.

22.   (a) C: $a_n^{1/n} \to \frac{1}{2}$.   (b) C: $a_n^{1/n} \to 0$. (c) C: $a_n^{1/n} \to 0$.

EXAMPLES 7   (PAGES 131-133)

1.   Show that $f'(0) = f'''(0) = f^{(5)}(0) = \dots = 0$.

2.   (a) See §174*.   (b) $(-1)^n n!/(1 + x)^{n+1}$. (c) Split into partial fractions and use (a) and (b). (d) $2^n n!/(3 - 2x)^{n+1}$.   (e) Write it as $1 - 1/(1+x)$.

(f) $(-1)^{n-1}(n-1)!/x^n$.   (g) Use Leibniz's theorem.   The answer is $2^{n-2}e^{2x}(4x^2+4nx+n^2-n)$.

3.   Notice that the (n+1)th and all higher derivatives of a polynomial of degree n are zero.   So the remainder term $R_{n+1}(x)$ is automatically zero.

4.   Use the method of §171*.   The polynomial is
$2 - 11(x-1) - 15(x-1)^2 - 3(x-1)^3 + (x-1)^4$.

5.   Suppose all the coefficients have positive sign. Then since x, x^2, ... all increase (through positive values) as x increases from 0, it follows that f is strictly increasing on $[0,\infty[$.   This contradicts its periodicity.

6.   Similar to §174*.

7.   Checking all the details of making the remainder tend to zero is rather wearisome.

8.   The series is $(1 - \frac{1}{2})^{-\frac{1}{2}}$.

9.   Introduce factors $2.4.6...(2n)$ above and below on the LHS.   Then pull out all the factors 2 in the denominator.

10.   Valid for $|x| < \frac{1}{2}$.

11.   The largest possible value is
$ab^{-2}(1+.05)(1-.10)^{-2} = 1.296\,ab^{-2}$ (approximately) by the binomial theorem.   So this is about 29.6% high.

12. $\sum_{n=0}^{\infty}(-1)^n\frac{(2x)^n}{(2n)!}$.   Differentiate the given

relation repeatedly.   (The function is actually $\cos 2x$.)

13.   Notice that $f''(0)$ can be got from the differential equation.

14.   Multiply out $(1-x)(1+x^3+x^6+...)$.

15.   (a) 2.   (b) $(3/2)^{2/3}$.   To find the binomial series from which the series come demands some enlightened guesswork: in (b) the 2.5.8 suggests $-\frac{2}{3} \cdot -\frac{5}{3} \cdot -\frac{8}{3}$.

16.   The series is $\sum_{n=0}^{\infty}\frac{(x-a)^n}{(1-a)^{n+1}}$.   It represents

the function on that part of the real line (regarded as a subset of R^2) that lies strictly inside the circle with centre $(a,0)$ and passing through $(1,0)$.

EXAMPLES 8   (PAGES 142-144)

1.   Proceed as in §179(ii).   Sum up the inequalities for $k = 1, 2, \ldots , (n-1)$.

2.   Apply the mean-value theorem to loglog x on $[2,3]$, $[3,4]$, $\ldots$, $[n-1,n]$, as in §179(ii) and then continue as in §179*.   Alternatively work with $\int_1^n 1/(x \log x)\, dx$ as in §179* from the outset.

3.   For the second part, sum up a set of inequalities given in the first part.   For the third part, take the exponential of the second part.

4.   C for $\alpha > 1$, D for $\alpha \leq 1$.

5.   C for $\alpha > 1$, D for $\alpha \leq 1$.

6.   Use the method of §180*.   Answers are log 3, $\log(3/2)$ and 0 respectively.

7.   Aim to use the method of §180* in each case. Some initial estimation may be needed.   For example, the sum in (a) lies between

$$\frac{1}{2n} + \frac{1}{2n+2} + \ldots + \frac{1}{4n-2} \quad \text{and} \quad \frac{1}{2n+2} + \frac{1}{2n+4} + \ldots + \frac{1}{4n}.$$

Take out the factor 2 in the denominators and apply §180*.   The answers are (a) ½ log 2 .   (b) (log 2)/4. (c) log 2.   (d) log 2.

8.   (a) Look at t_{3n} .   (b) Look at t_{3n} . (c) Look at t_{5n} .   In each case use the result of §180.

9.   Look first at $t_{n(p+q)}$ and use §180.

10.   Take it as $\left(\sum\limits_{r=0}^{\infty} x^r \right) \left(\sum\limits_{s=0}^{\infty} x^s \right)$, i.e. with

different suffices as in §187.

11.   You will need to write $1/(r(n-r))$ in partial fractions.

13.   Notice first that $1+2+3+ \ldots +n = \frac{1}{2}n(n+1)$. So $t_{n(n+1)} = 0$.   We must also check that other partial sums do not stray far from 0.   To do this write out the (n+1)th batch of positive terms explicitly and use the result of §180 to show it tends to zero as $n \to \infty$.

14.   Compare Ex.13 above.   Here show that the (n+1)th batch of positive terms tends to a non-zero limit as $n \to \infty$.

EXAMPLES 9 (PAGES 153,154)

1. (a) $R = 1$; C at ± 1. (b) $R = \frac{1}{2}$; C at $\pm\frac{1}{2}$.
(c) $R = \frac{1}{2}$; C at $\frac{1}{2}$, D at $-\frac{1}{2}$.

2. (a) $1/27$. (b) 1. (c) infinite. (d) $\frac{1}{2}$.
(e) 0. (f) $\frac{1}{2}$. (g) 1.

3. $\left|a_{n+1}/a_n\right| = 9(n+1)^2/((3n+1)(3n+2))$ when $|z| = 27$.
Use §152.

4. Write $|z-1| < |z+1|$ as $|z-1| < |z-(-1)|$.
This says that the distance of z from 1 is less than
the distance of z from -1.

6. To show that $R = 1/\sqrt{6}$ show first that the
series diverges when $z = 1/\sqrt{6}$. Then show it converges
for $|z| < 1/\sqrt{6}$. (It may help to take odd and even
powers as two separate series.)

7. $R = 1/\sqrt{2}$.

8. Use the ratio test.

9. $p = 2$, $q = -1$.

10. (a) a. (b) $\frac{1}{2}a^2$. (c) 1. (d) $-\alpha a^2$. Use
the power series for sine and cosine from §216 for the
first three. In (d) expand with the binomial theorem.

11. Expand the left using the binomial theorem.

12. Differentiate, multiply through by x,
differentiate again, etc.

13. Use the method of §196.

14. Suppose that after the coefficient a_N all
other coefficients are positive. Find the $(N+1)$th
derivative of F, notice that it is periodic and apply
Ex.7.5.

15. Notice that $\left|b_n z^n\right| \leq \left|a_n z^n\right|$ for all $n \geq 0$.

EXAMPLES 10 (PAGES 175-181)

1. Find the turning point.

2. By §209, $n/\log n \to \infty$ as $n \to \infty$. So
$n/\log n > 1$ for all $n > X$. So $n > \log n$ for all
$n > X$. So $1/\log n > 1/n$ for all $n > X$. Since
$\sum 1/n$ diverges, so also does $\sum 1/\log n$ by comparison.

A similar argument shows that $\sum 1/(\log n)^k$ diverges
for every fixed $k\varepsilon N$.

3. (a) $\frac{1}{2}(b^2 - a^2)$. (b) 1. (c) 0.

4. For the first part, see §134(ii). For the

EXAMPLES 10 (PAGES 175-181) (CONTINUED)

second part, raise the inequality to the power k and use comparison. Answer is C for $k > 1$, D for $k \leq 1$.

5. Put $x = 1/(2n+1)$ in §213(ii).

6. (a) Use the method of §214. Answer $= e$.
(b) Divide above and below by n^2. Use the method of §214. Answer $= 1$. (c) See §211*. Answer $= 0$.
(d) $n^{1/\log n} = e$ for all $n \geq 2$. (e) $\log(a/b)$.
(f) 0. (g) Write it as $\exp(n^2 \log 5 - n \log n)$ and use the hierarchy of §209. It tends to infinity.

8. See Ex.6.16. Answer $= -\frac{1}{2}\log 2 + i\frac{\pi}{4}$.

9. (a) D: terms exceed $1/n$ for $n \geq 3$.
(b) D: integral test. (c) C: integral test.
(d) C: Cauchy's root test.

10. Use §123.

11. Let $x = \log_a b$. So $a^x = b$ etc.

12. Apply the mean-value theorem to $\log_{10} y$ on the interval $[125,128]$. Then use $128 = 2^7$ and $125 = 1000/2^3$.

13. Take 3^{10} as 5.9049×10^4. To get 3^{100} find 5.9049^{10} on the calculator and multiply it by 10^{40}. Continue in this way to split off a power of 10.

14. Write $2^x = e^{x \log 2}$.

15. Use the method of §178.

16. Take $\log(1 - x^3) - \log(1 - x)$ and expand with §212.

17. Show that $x - \log(1 + x)$ is increasing on $[0,1]$ in the same method as in §131.

18. Reduce each tail to a geometric series.

19. Use identities for $\sin(a + b)$ etc. Derive the one for $\tan 3x$ from the other two.

20. Use the method of §223. In each hyperbolic identity the sign of $\tanh^2 z$ is the opposite of that of $\tan^2 z$. Otherwise no alterations.

21. Use $\tan^{-1}a - \tan^{-1}b = \tan^{-1}\left(\dfrac{a - b}{1 + ab}\right)$.

24. $x^{12}e^{-x} \to 0$ as $x \to \infty$. So there exists M such that $x^{12}e^{-x} < 1$ for all $x \geq M$.

25. $R = 4/e^2$, by taking the ratio of consecutive terms and using the result of §214. It is divergent for $z = R$, since then $a_n \sim 1/\sqrt{(4\pi n)}$ by Stirling.

26. Limits are (a) ∞, (b) 1, (c) ∞, (d) 1, (e) 1, (f) 1, (g) ∞. For (a), (b), (c), (e) use the fact that $a^x = e^{x\log a}$ as in §214. For (d), notice that

$$\left|1 + \frac{i}{n}\right| = \left(1 + \frac{1}{n^2}\right)^{\frac{1}{2}}.$$

For (f), first write $\log(n+1)$ as $\log n + \log\left(1 + \frac{1}{n}\right)$.

27. (a) 1. (b) 0.

28. For the first part use the intermediate value theorem. Then let $f(x) = \tan^{-1}(e^x)$, and aim to use §135(i). The inequality of the means (§39) shows that $0 < f'(x) \le \frac{1}{2}$. §135(ii) is a model for the last part.

29. Very similar to Ex.28 but just estimate the derivative rather than use the inequality of the means.

30. Let $f(x) = \tan^{-1}(e^{3x})$. The method of the two preceding examples fails because $f'(x)$ can exceed 1. Notice that $-\frac{1}{2}\pi < a_n < \frac{1}{2}\pi$ ($n \ge 2$). Use the mean-value theorem on $f(a_n) - f(a_{n-1})$ to get the monotonicity.

31. Differentiate the first series to get the series for $\tan x$.

32. Write $(n+1)^2 = n(n-1) + 3n + 1$ and hence split the given series into three series.

33. Take $\left|1 + \frac{z}{n}\right|$ first.

34. Prove that $e^x/e^{(\log x)^2} \to \infty$ and $e^{(\log x)^2}/x^k \to \infty$ as $x \to \infty$. For the second part of this write $x^k = e^{k\log x}$.

35. Choose n such that $\log(n+1)$, $\log(n+2)$, ... , $\log 2n$ all lie between $(2k+\frac{1}{4})\pi$ and $(2k+\frac{3}{4})\pi$ for some positive integer k. Then if s_n denotes the nth partial sum of the series, $s_{2n}-s_n$ exceeds

$$\frac{1}{\sqrt{2}}\left(\frac{1}{n+1} + \frac{1}{n+2} + \cdots + \frac{1}{2n}\right).$$

Then use §180*.

37. (a) No. (b) Yes. It may help to use the mean-value theorem to estimate $\log x - \log y$.

38. You will need the value of $\binom{2n}{0} + \binom{2n}{2} + \cdots + \binom{2n}{2n}$ from algebra books.

39. For $y > 0$, constant is $\frac{1}{2}\pi$. For $y < 0$, constant is $-\frac{1}{2}\pi$.

41. Find the real and imaginary parts of

$\sum\limits_{n=1}^{\infty} a^n e^{in\alpha}$, using the formula for the sum of a geometric

series.

42. $\sqrt{x} \tan^{-1}\sqrt{x} - \frac{1}{2}\log(1+x)$.

43.   Use §213(ii) together with the series for
$\tan^{-1}x$ in §220.

44. $2\pi i$.

45.   Use $a^x = e^{x \log a}$.   For (a), answer = 1.
For (b), the limit is infinite from the behaviour in
Ex.2.27.

46.   First find $\left|1 + \dfrac{i}{\sqrt{n}}\right|$.

47.   Take the logarithm of each side.   For the
"deduce", notice that $\log\log x > 2$ for all $x > N$, say.

48. $e^{\sqrt{(\log x)}}$.

49.   First write $a^{1/n} = \exp\left(\dfrac{\log a}{n}\right)$ and expand

this as a power series.   Then use the method used in
§214.   The answer is $\sqrt{a}$.

50.   (a) ∞.   (b) 0.   (c) Put $x = 1/y$ and let
$y \to \infty$.   Answer = e.

51.   See §131.

52.   See §131.

EXAMPLES 11   (PAGES 203-205)

1.   Show that $L(f:D) = 1$ and $U(f:D) = 2$ for
every dissection D.   For the example see Ex.3.8.

2.   Denote $1/f$ by g.   Then $|g(x) - g(y)| \leq$
$\dfrac{1}{k^2}|f(x) - f(y)|$.   Look at $\Delta(g:D)$.

3.   Since $\sin(\pi-x) = \sin x$ and $\cos(\pi-x) = -\cos x$
for all $x \in [0,\frac{1}{2}\pi]$, it follows that an odd power of the
cosine in a product $\sin^m x \cos^n x$ produces as much
area below the x-axis between $\frac{1}{2}\pi$ and π as positive
area between 0 and $\frac{1}{2}\pi$, so that the second integral
given is zero.   On the other hand, an even power of the
cosine produces the same area above the x-axis
in both intervals, so that the first integral is twice
the integral from 0 to $\frac{1}{2}\pi$.
    The parity of the index of the sine is irrelevant as
the sine is non-negative on the whole of $[0,\pi]$.

4.   (i)   Integrate $1/x$ on $[2,3]$ or integrate $1/(x+2)$ on $[0,1]$.   (ii)   Integrate $1/x^2$ on $[2,3]$ or integrate $1/(x+2)^2$ on $[0,1]$. (iii)   Integrate $1/(1+x^2)$ on $[0,1]$.

5.   Look at $\int_0^1 x^p\, dx$.

6.   Look at $\int_0^1 \log(1+x)\, dx$ or $\int_1^2 \log x\, dx$. For the "deduce", combine all the logarithms in the first part.   Then take the exponential of each side.

7.   Use $\int_0^1 x \log x\, dx$.   For the "deduce", use $k \log a = \log a^k$ and combine the logarithms.   You will also need the sum of $(1 + 2 + \ldots + n)$ from Ex.1.14.

8.   (a) Put $x = 3y$.   (b) Double the integral from 0 to ∞.   (c) Put $x^4 = y$.   (d) It is in form (iii) in §254.   (e) Put $y = \log x$.   (f) Put $y = 8x^3$.

12.   Similar to Ex.6 above.

14.   In the last part, take real and imaginary parts at the end.

EXAMPLES 12   (PAGES 214-215)

2. $R = 1/e$.   At $z = R$, $a_n \sim 1/\sqrt{(2\pi n)}$.

3.   Show $f(n+1)/f(n) = 1 + O(1/n^2)$.   Apply §268.

4.   Show $f(n+1)/f(n) = 1 + O(1/n^2)$.   Apply §268.

5.   Similar to §275.

6.   Use $2\cos^2 n\theta = 1 + \cos 2n\theta$ and the fact that the sum of a divergent series and a convergent series is divergent.   For the last part, use the fact that $0 \leq \cos^2 n\theta \leq |\cos n\theta|$.

7.   Try $\alpha = \pi/2$, π, 2π, $\pi/4$, $\pi/3$, etc.   It may help to notice that, for all odd $n \in N$, $n^2 \equiv 1 \pmod 4$.

8.   If $\alpha = p/q$, then $z^{n!} = 1$ for all $n \geq q$.   So the terms eventually just become those of $\sum 1/n$.

9.   Radius of convergence is 1.   It is absolutely convergent for $|z| = 1$ by Gauss's test.

10.   For $|z| = 1$, $|a_{n+1}/a_n| = 1 - \dfrac{\alpha+1}{n} + O(1/n^2)$. So, if $\alpha > 0$, the series is absolutely convergent. If $\alpha < -1$, then $|a_{n+1}/a_n| > 1$ for large n and so the series diverges by §152.   For $-1 < \alpha < 0$, use Dirichlet together with §272 to see that the series diverges at $z = -1$ and converges at all other points with $|z| = 1$.

EXAMPLES 13   (PAGES 216-218)

1.   For $\alpha^{1/7}$ take $x_{n+1} = (\alpha x_n)^{1/8}$.   For $\alpha^{1/5}$ take $x_{n+1} = (\alpha^3 x_n)^{1/16}$.   For $\alpha^{1/p}$ use number theory to show that, for every odd prime p, there exists a positive integer n (with $n \leq (p-1)$) such that $2^n \equiv 1 \pmod p$.   Proceed from there.

2.   Compare Ex.4.19.

3. $L - \varepsilon < a_{k+1}/a_k < L + \varepsilon$ for all $k \geq N$.   Take a product of these inequalities for $k = N, N+1, \ldots, n$. Then take the nth root.

5.   Show that f is increasing and bounded above. Rule out the possibility of a non-zero limit using §133.

7.   A sequence wich tends to a limit is bounded above by K, say, by §49.   So
$$a_3 + a_4 \leq K/2^2, \quad a_5 + a_6 + a_7 + a_8 \leq K/4^2 \quad \text{etc.}$$
Then add up.

8.   Use the method of §179.

9.   For the last part, use the RHS of (2) for n and the LHS of (2) for $2n$.   (It is possible to get a better inequality in the last part by another method.)

11.   (i) See Ex.8.7.
$$(ii) \quad t_{3n} = \left(1 + \frac{1}{5} - \frac{1}{3}\right) + \ldots + \left(\frac{1}{8n-7} + \frac{1}{8n-3} - \frac{1}{4n-1}\right)$$
$$= \left(1 - \frac{1}{3} + \frac{1}{5} - \frac{1}{7} + \ldots + \frac{1}{4n-3} - \frac{1}{4n-1}\right)$$
$$+ \left(\frac{1}{4n+1} + \frac{1}{4n+5} + \ldots + \frac{1}{8n-3}\right).$$

The sum is $\frac{\pi}{4} + \frac{\log 2}{4}$.

13.   Use $|z - \bar{z}| = |2i \operatorname{Im} z| \leq 2|z|$.

14.   Not entirely by chance: both have a connection with the Pythagorean triple (119, 120, 169).   If (a, b, c), where $a, b, c \in \mathbb{N}$, is a Pythagorean triple with c as hypotenuse, then there is a rational solution x of the equation $2 \tan^{-1} x = \tan^{-1}(a/b)$.   In Machin's formula a/b is taken as 120/119.   In the formula in the example a/b is taken as 119/120.
You can construct
$$2 \tan^{-1} \frac{2}{5} + \tan^{-1} \frac{1}{41} = \frac{\pi}{4} \quad \text{and} \quad 2 \tan^{-1} \frac{3}{7} - \tan^{-1} \frac{1}{41} = \frac{\pi}{4}.$$

(Another Pythagorean triple where the two short sides differ by 1 is (696, 697, 985).)

15.   The sum is $\frac{\pi}{4}$ - log 2.

16.   See Ex.10.43.

18.   Write $n^{\log n} = e^{(\log n)^2}$ and write $a^n = e^{n \log a}$. Also log a < 0. Alternatively use the method of §65.

19.   Use the method of §65. Limit is 0.

20.   Note that $n^{-i} = \cos(\log n) - i \sin(\log n)$ from §228. Then use Ex.10.35 to show that the imaginary part of $\sum 1/n^{1+i}$ diverges.

21.   Integrate $x^{\alpha}(1-x)^{\beta}$ on [0,1].

SOME MACLAURIN SERIES

Function	Maclaurin series	valid on
e^x	$1 + \dfrac{x}{1!} + \dfrac{x^2}{2!} + \dfrac{x^3}{3!} + \ldots$	R
e^{-x}	$1 - \dfrac{x}{1!} + \dfrac{x^2}{2!} - \dfrac{x^3}{3!} + \ldots$	R
$\cos x$	$1 - \dfrac{x^2}{2!} + \dfrac{x^4}{4!} - \ldots$	R
$\sin x$	$x - \dfrac{x^3}{3!} + \dfrac{x^5}{5!} - \ldots$	R
$\cosh x$	$1 + \dfrac{x^2}{2!} + \dfrac{x^4}{4!} + \ldots$	R
$\sinh x$	$x + \dfrac{x^3}{3!} + \dfrac{x^5}{5!} + \ldots$	R
$\tan^{-1} x$	$x - \dfrac{x^3}{3} + \dfrac{x^5}{5} - \ldots$	$[-1,1]$
$\log(1 + x)$	$x - \dfrac{x^2}{2} + \dfrac{x^3}{3} - \ldots$	$]-1,1]$
$\log(1 - x)$	$-x - \dfrac{x^2}{2} - \dfrac{x^3}{3} - \ldots$	$[-1,1[$
$(1 + x)^k$	$1 + \dfrac{k}{1!} x + \dfrac{k(k-1)}{2!} x^2 + \ldots$	See §175.
$1/(1 - x)$	$1 + x + x^2 + x^3 + \ldots$	$]-1,1[$
$1/(1 + x)$	$1 - x + x^2 - x^3 + \ldots$	$]-1,1[$
$1/(1 - x)^2$	$1 + 2x + 3x^2 + 4x^3 + \ldots$	$]-1,1[$

INDEX

All numbers refer to sections <u>not</u> to pages. Certain symbols are listed separately at the end.

SOME NOTATION

———————————

ALSO AVAILABLE:

ADVANCED CALCULUS FOR ENGINEERING AND SCIENCE STUDENTS
Ian S. Murphy, 1984.
Arklay Publishers, 64 Murray Place, Stirling.

Contents: Double and Triple Integration, Beta and Gamma
Functions, Differential Equations, Laplace Transforms,
Partial Differentiation, Errors and Exact Differentials,
Vector Calculus, Line and Surface Integrals, Fourier
Series, Maxima and Minima of Functions of Several
Variables including Lagrange Multipliers.

———————————

Azman

0171 790 1875 Lon.

01702 584874 S.E.